90

Stanley P. Chase

1910.

# SHELLEY'S LITERARY AND PHILOSOPHICAL CRITICISM

EDITED WITH AN INTRODUCTION

BY

JOHN SHAWCROSS

LONDON

HENRY FROWDE

1909

OXFORD: HORACE HART
PRINTER TO THE UNIVERSITY

# INTRODUCTION

THE present edition of Shelley's prose works comprises all his later critical and speculative writings[1] (with the exception of the prefaces to the longer poems) and a selection from his letters, illustrating more particularly his literary and artistic criticism. The youthful romances and the various political pamphlets, as well as the lengthy notes to *Queen Mab*, have been omitted from the volume, the primary aim of which is to exhibit Shelley's maturer genius on its critical and philosophical side.

This principle of selection has necessarily confined the edition within somewhat narrow limits. Except under the pressure of some great public occasion (and then only in his earlier years), Shelley was not readily moved to sustained utterance in prose. At first sight this may appear strange, for he was by no means averse from those interests which find in prose their natural vehicle of expression. Apart from his 'passion for reforming the world', Shelley had also a passion for speculating upon it and unravelling its meaning. Of the first he has left abundant record in the reform pamphlets of his youth; yet after 1812, with the exception of the Marlow pamphlet, he wrote nothing with a directly practical aim. His love of

[1] I refer to published writings only.

speculation grew stronger rather than weaker, as his years increased, and seems, to some extent, to have tempered his political zeal; in the *Philosophical View of Reform*, written in 1818, the two tendencies blend and are reconciled, and the spirit of this paper may perhaps be taken as typical of Shelley's maturer attitude to political and social problems. Yet in spite of this ardour for theorizing, Shelley has left behind him in writing only the veriest fragment of his speculations. Nearly all that he has written of this nature is confined within a period of a few months, the latter part of the year 1815; and it covers less than a hundred pages of print. After these attempts were abandoned, Shelley produced nothing purely speculative[1] until the end of 1820, when Peacock's attack on poetry and poets 'excited his polemical faculties', and drew from him his finest sustained effort in prose writing. But of this work the first portion only was completed: as Shelley's indignation subsided, the impulse to write subsided likewise, and although at the end of this year we find him still contemplating the completion of his apology, there is no evidence that he actually finished it. The disquieting reception of *Epipsychidion*, which even the συνετοί, as Shelley wrote, wholly failed to comprehend, moved him to project a *Symposium*, 'in which all these misunderstandings should be set right'; but in the end he was content to leave things as they were. Nor was the more ambitious design, to which he alludes in the preface

[1] The prefaces to the longer poems contain much valuable speculation, but this is incidental to their main purpose.

to *Prometheus Bound*, of 'a systematical history of . . . the genuine elements of human society', carried into execution during the remaining years of his life.[1]

Shelley's work in prose, then, seems at first sight, if judged by its bulk alone, surprisingly slight and insignificant. But the cause of its fragmentary nature is perhaps not far to seek. To direct intervention in the political and social questions of the day Shelley grew less and less inclined, as his years increased and his self-knowledge ripened.[2] As regards his speculations, moral or metaphysical, Shelley was, we have to remember, an uncompromising disciple of Godwin, and the majority of his longer poems are impregnated with Godwin's ideas. Now to reproduce those ideas in poetic guise might naturally seem well worth the while to Shelley, and it is true that, Shelley being the poet, the attempt was more than justified; but to systematize them in a prose work, unless by way of illumination or correction, is on the face of it a superfluous task, and so it probably appeared to Shelley himself, when he so soon abandoned the attempt. Here, then, we have an explanation of the fragments of 1815: but we have not yet understood why the *Defence*, which lay outside Godwin's sphere, or the *Philosophical Review of Reform*, in which Shelley is beginning to emancipate

[1] Unless indeed Shelley's unpublished MSS. contain anything of the kind. But it is presumable that any important remains would have seen the light ere this.

[2] Cp. Letter to Horatio Smith (1822), 'I once thought to study these affairs, and to write or act in them. I am glad my good genius said *refrain*.'

himself from his master, are unfinished, nor are we justified in concluding that even had Shelley lived long enough to render that emancipation complete, his contribution to prose literature would still have been a scanty one. Yet this conclusion is probably true, and this for the simple reason that, in spite of his own assertion,[1] Shelley was, by genius and inclination, a poet, and that the true energies of his nature were fully claimed by his poetic vocation. Should his achievement in the *Defence* be quoted against this argument, it must be remembered that the *Defence* stands to some extent apart from the rest of his work in prose. Not only is it inspired, like nearly all his longer writings, by the sense of high injustice done to a noble cause, but the subject is sufficiently ideal to admit of an imaginative treatment; and, were we to measure it by his own criterion of poetry, we could hardly deny it the title of a poem. The existence of the *Defence* is therefore no guarantee that Shelley, had he lived longer, would have succeeded in a more speculative type of writing, or even that he would have proved himself further in this field.

Nor would the absence of such a development of his genius, it may be added, have been a thing for posterity to regret; any more than, as concerns his actual life, it is regrettable that Shelley never devoted

[1] Letter to Peacock, Jan. 26, 1819: 'I consider poetry very subordinate to moral and political science, and, if I were well, I would aspire to the latter.' The force of this statement, which in itself is not perhaps to be taken too seriously, is considerably qualified by the words which follow.

his time and energies to a systematic exposition of his ideas. For it is scarcely possible that he could have prosecuted this object without a corresponding loss of imaginative power ; and what Goethe declared of Schiller would have held equally true of Shelley, that 'through his concentration on philosophy he came to place the Idea above nature, nay, indeed, to destroy nature thro' the Idea'. From such a reproach, indeed, Shelley is not, as it is, wholly to be exempted ; had he followed Schiller's example, it is certain that his inherent defects as a poet, instead of yielding to his maturer artistic insight, would have obscured that insight more and more, and stamped his noblest efforts with the mark of imperfection.

That his habit of reflection did not, during his actual life, thus come to dominate his mind is therefore a matter for satisfaction, and this not to lovers of his poetry alone. If Schiller's speculations on poetry impaired his actual poetic work, they at least produced results of unquestioned value. But it is more than doubtful if there would have been a like compensation in Shelley's case. For Shelley was a daring rather than an original, a subtle rather than a scientific thinker. His prose writings are more remarkable for clarity than for depth of thought, for logical precision in the following out of a particular line of argument rather than for method or self-consistency in the presentation of a complete theory or point of view. Scientific as he was in his passion for law (in this the poet and the scientist in him meet), he lacked, or at least was slow to cultivate, the strictly scientific habit of mind, the habit of rigid accommodation of preconception to fact, of generalization to experience; his high respect for reason often led him, as we shall see, to conclusions far from reasonable. Hence we shall be disappointed if we expect from Shelley's prose writings anything approaching a unified system, or even a definite contribution to mental or moral science. Their true value depends rather on his qualities as a poet than as a thinker: he is seen at his best not when he is expounding a theory in the light of sober reason, but on occasions which have stirred his feelings and his imagination; when he is upholding some cherished conviction, depicting some deep emotion, or delineating in words some beauty of nature or art.

But although Shelley's speculations have no high scientific import; although, further, it is difficult to piece these fragments into a coherent scheme of thought, and call this Shelley's philosophy; yet his speculative writings have their value to the student, for they illustrate his genius and characteristic temper of mind. And the attempt is perhaps justified to exhibit these writings, if not as portions of a coherent scheme of thought, yet at least as dominated by the same attitude to life and reality. To say that Shelley's mind moves naturally in the region of the ideal, that the concrete and the particular are antipathetic to his genius, is to say nothing new; it is perhaps to re-iterate an opinion which, often too lightly accepted, has given birth to one-sided estimates of his work. Yet it cannot be denied that the prejudice has a root of truth, which deserves to be brought to the surface. That Shelley's affections are frail and ethereal, his vision wholly averted from the things of earth, are extravagant assertions which it is easy to refute; this need not, however, preclude us from attributing to his genius a predilection for what is simple, definite, and abstract. But it is necessary to be careful in our use of terms. This predilection in Shelley differs from the logician's delight, often arid and unimaginative, in abstract conceptions. It is from his keen sense of the obscurity, the complexity, and changefulness of the material world, where 'nothing endures but mutability', and where so much seems accidental, aimless, and superfluous, that Shelley's longing for a world of permanent forms, and his faith in its existence, are sprung. Hence it is his constant endeavour to view the world of sense in the light of this immutable reality; and his love of Nature is fastened upon that which in her manifestations seems actually or symbolically most akin to the permanent and immaterial. This attitude is reflected, in his poetry of nature, in the comparative scarcity of those concrete images which detain and yet stimulate the imagination, and in the predominance of the simple impressions of colour, light, and motion.[1] That his poetry fails on this account as an interpretation of Nature, or as a revelation of her beauty, is not to be inferred; for the poet employs

[1] In actual landscape Shelley seems to have preferred clear and sharply defined features. He speaks disparagingly of the scenery of the Apennines as 'broad and undetermined'—'the imagination,' he says, 'cannot find a home in it.' But this is no wise inconsistent with the comparative absence in his nature-poetry of distinct images.

natural imagery not for its own sake, but for the purpose of expressing and arousing certain definite emotions, and in this Shelley does not fail of success. In his love-poetry, again, the passion, however vital and genuine, seems often too transcendent and generalized: it is not so much the communion of one individual soul with another that we witness, as the poet's own longing to consummate the greatest of life's experiences. The particular, in fact, is sacrificed to the universal, but in compensation there is an ideal and inspired beauty in his lyrics which few poets have rivalled. The same tendency is apparent where his imagination is at work upon the materials of human nature and life. To it may be referred the allegorical treatment of the problems of life in his longer poems; and in his earlier poetry, at least, the idea is prior to and determines its embodiment, so that Shelley runs dangerously near to being didactic. The defects of his character-drawing illustrate the same idiosyncrasy. To expect complexity in the heroes of poems professedly allegorical is unreasonable: but even in *The Cenci* the central characters, true dramatic creations as they are, exhibit in their simplicity and directness Shelley's aversion from the indeterminateness and intricacy of life.

But it is in Shelley's prose-writings and the theories they embodied that we are primarily concerned to discover this characteristic of his genius. In philosophy Shelley was in early days a materialist, and in materialistic writings he met the doctrine of Necessity, his attachment to which seems to have survived his materialism. This latter doctrine, though at first sight it appears to associate itself with an undue respect for the concrete, is in reality highly idealistic and abstract, inasmuch as, though it professes to start from nature, it ignores the fundamental character of natural things—that is, their complexity. No doubt it was the apparent simplicity of the doctrine which first attracted Shelley to it; but in later years he saw its fallacies, and was 'discontented with such a view of things as it afforded'. Nor could 'the popular philosophy of mind or matter' (Shelley's designation, apparently, for the common-sense school of thought) be more congenial to his idealizing temper. In this it is instructive to compare him with Coleridge, who, in spite of the familiar accusations of cloudiness, did yet in his philosophy honestly attempt to deal with and justify the common-sense attitude to the concrete world. Shelley, the lover of clarity, is too apt to ignore the actual, and thus his philosophy, like his poetry, becomes 'dark with excessive bright'. It is this temper in him which, when he has shaken off materialism, drives him into the arms of Berkeley, and even beyond Berkeley to Hume. In this development, no less than in that of his social and political theories, the influence of Godwin must not be ignored. In Godwin's writings Shelley found the 'intellectual system' stated in its barest terms. Thus in 1815 (in the *Essay on Life*) we find Shelley declaring, 'I am one of those who am (*sic*) unable to refuse my assent to the conclusions of those philosophers who assert that nothing exists save as it is perceived'. And in the speculations on metaphysics he writes: 'It is an axiom that we can think of nothing which we have not perceived. When I say we can think of nothing, I mean, we can imagine nothing, we can reason of nothing.' But not only is mind thus dependent, according to Shelley, for its activity upon the presentations of sense: it appears even to be identified with these objects in their various combinations, and thus to disappear as an active originating force. 'The elements of the human mind,' he writes in this same essay, 'being abstracted, sensation and imagination cease. Mind cannot be considered pure.' And not only would Shelley explain mind, as a separate force, away: even the varied materials of consciousness, 'thoughts or ideas or notions,' he assimilates to one another, denying their essential variety, and conceding reality only to that which they have in common, their quality as thoughts. In this Shelley is a disciple of Godwin; but he is also true to his own passion for abstraction and simplification. For he thus erects for himself 'a conception of Nature inexpressibly more magnificent, simple, and true, than accords with the ordinary systems of complicated and partial consideration': and such a conception it is that he delights to contemplate. Had he, however, been more self-critical, or pursued his researches further, he must surely have found this view of things, for all its simplicity, as little fitted as that of the materialists to satisfy 'a being of high aspiration, looking before and after', a being whose spirit was 'at enmity with nothing and dissolution'; he must have found it, in fact, wholly irreconcilable with his own idealism.

In Shelley's moral philosophy the most significant fact is, perhaps, his attachment to the doctrine of necessity, which interested him perhaps more in its ethical or religious than in its psychological aspect. At first sight nothing would appear more repellent to a being of his keen sensibility and vivid imagination than this grim and mournful doctrine, which seems to sound the knell of all individual faith and endeavour. But here again we must not be at the mercy of terms; necessity to the psychologist may stand for one thing, and to the poet for something widely different. To Shelley it revealed itself as an aspect of that permanent and changeless reality, fixed far above the flux of time, which was the object of his lifelong aspiration. His necessarianism is thus the earliest form of his so-called Platonism. To the author of *Queen Mab* Necessity is 'a goddess above the tide of human passion and the varying and multitudinous shapes of life'. Later, the goddess descends to earth: in *Laon and Cythna* it is as the immutable principle of cause and effect in the sphere of human action, the power

> whose sightless strength for ever
> Evil with evil, good with good must bind,

that necessity presents itself to the poet's imagination. But in either case it is the satisfying sense of something simple and inviolable that appeals to Shelley in this conception, and perhaps helps him to ignore the unpleasing implications of the necessarian doctrine.

Shelley's attitude to moral questions is largely that

of Godwin ; and no doubt the attraction of Godwin's scheme for Shelley lay in its apparent simplicity and reasonableness. Godwin sought to simplify ethical problems by converting them into intellectual. The unguided feelings are, in his conception, the source of all moral error : for in them originates the fundamental frailty of human nature, its tendency to attach itself to the particular environment into which fate has thrown it. This tendency, considered in itself, is wholly unreasonable, and all conduct based upon it is to be condemned as wrong. Our attachments must, in fact, spring solely from 'a rational perception of merit'. 'All attachments to individuals, except in proportion to their merits, are unjust.' Godwin would indeed wholly eliminate the emotions from the sphere of moral action. 'Man is a moral being no further than he is capable of connecting and comparing ideas.' 'To perceive that I ought is to perceive that a certain thing is preferable.' Virtuous preference is preference for an object from a clear perception of its value.

This ambition of Godwin to free human action and life from all that is seemingly incidental and unmeaning was no doubt laudable ; and his conviction that only through the guidance of reason could such freedom be attained is not difficult to understand. But he erred in a double sense, and stands convicted, apostle of reason as he was, of high unreasonableness. He seeks to eliminate the element of unreasoning feeling, and yet bases the whole of human conduct on one fundamental emotion, the desire to promote the general well-being : a desire which is itself wholly unreasonable, unless we understand by reason something very different from what Godwin understood by it. In order to escape this dilemma he confounds intellectual perception with moral choice, and thus his argument becomes equivocal throughout. Man is emphatically not 'governed in his actions purely by an unimpassioned survey of the nature of things'. His primal instinct of benevolence is an emotion, and emotion must therefore be allowed a voice in all his decisions and attachments. Moreover, man by his nature is individual, and hence attaches himself inevitably to what is individually and, as it appears, accidentally related to himself. Godwin, in fact, embraced in the name of reason a doctrine which is highly irrational, because it enters into blind conflict with incontrovertible fact, with the unalterable nature of things.

It is, then, no doubt in Shelley's aversion from the concrete and the complex that we must seek an explanation of the remarkable influence which Godwin's writings exercised upon him. Here Shelley saw propounded a scheme of morals which, while apparently recognizing the general claims of the feelings, sought to avoid the inconveniences arising from their habit of attaching themselves to particulars. But while Shelley conceives morality to have in so far an intellectual source, as its strength in the individual is proportionate to his knowledge and the expansion of his sympathies, he sees clearly the unreasonableness of attempting to explain rationally, that is, to give a reason for, the primary impulse of virtue. 'If a person persists to inquire,' he writes, 'why he ought to promote the happiness of mankind, he demands a mathematical or metaphysical reason for a moral action. The absurdity of this scepticism is more apparent, but not less real, than the exacting a moral reason for a mathematical or metaphysical fact.' Yet this primary impulse, although it admits of no rational explanation, is still, according to Shelley, essentially rational in the mode of its operation. 'We are led by our benevolent propensities to regard every human being indifferently with whom we come into contact.' Hence the benevolent impulse must often clash with the particular affections, which are determined by chance and by irrational considerations —above all, by the force of association; and where it comes into conflict with them, it is right, Shelley would say, that the primary impulse should conquer. In his insistence upon the absolute impartiality of benevolence, Shelley, it may be noted, is even more thoroughgoing than his teacher; for while Godwin allows reasonable grounds of preference to determine the direction of our impulses, which he calls 'attachment to individuals in proportion to their merits', Shelley concedes no ground of attachment save that quality which all beings possess in common, that is, their human nature. Moreover, in Shelley's belief this extension of the range of the affections implies no diminution of their intensity. 'You ought to love all mankind, nay every individual of mankind,' he writes: 'you ought not to love the individuals of your domestic circle less, but to love those beyond it more.' In these words we see expressed that faith in the infinite capacity of Love, to which Shelley afterwards gave poetic utterance in *Epipsychidion*.

To emphasize unduly what is unscientific or impracticable in Shelley's scheme of morals would be scarcely fair to its author. The precepts of the idealist must not be taken too literally : it is sufficient if their spirit, adopted as a basis of action, has beneficial results for the individual and the community. Shelley himself was far from dreaming of the immediate practicability of his ideas. 'It is necessary,' he writes, 'that universal benevolence should supersede the regulations of precedent and prescription, before these regulations can safely be abolished.' So far at least as Shelley's war upon existing institutions, private and public, was directed against the transient forms in which the limitations of human nature are embodied, and not against those limitations themselves, it is not only justified, but just and laudable. And of those limitations themselves who shall say to what extent they are absolute and unmodifiable ? It seems impossible to deny that the changes which moral conceptions have undergone since the days of Shelley and Godwin accord to no small extent with the spirit of their teaching.

It is difficult, however, to derive from Shelley's fragmentary utterances on moral questions (the chief of which is not merely unfinished, but even, as a fragment, incomplete), a satisfying conception of his views. This difficulty is increased by the seeming inconsistency of many of his statements. Thus, in his desire to prove that virtue is a matter of the head rather than of the heart, Shelley goes so far as to



24x

speak of it as ‘ entirely a refinement of civilized life ’, although it is to this same civilized life that he attributes that misdirection of the benevolent impulses to correct which is the first object of reform! Again (in his essay on *The Punishment of Death*), Shelley expressly refers the development of the various passions, good and bad, to the power of association, that is, to an essentially irrational mental procedure, which he himself designates as ‘ the principle of the mind overshooting its mark ’. Such contradictions do not in themselves invalidate Shelley's main position; but they raise the presumption that he never clearly elucidated that position in his own mind, a presumption fortified by the unfinished state of these reflections.

As in his moral, so also in his social philosophy, Shelley is a follower of Godwin. Indeed, their scheme hardly admits of a distinction between the two sciences; for it merges private and public morality in one. Here again Shelley's impatience of the actual and the concrete manifests itself. It would indeed be hopelessly wide of the mark to stigmatize him as an anarchist in the literal sense of the word. As a poet and artist, Shelley is essentially a lover of order; in order he sees the principle of beauty, whether expressed in sensuous form or civic institution. It is against defective institutions that his attack on society is aimed—defective, because they fail to reflect outwardly the inner moral law. One can understand, and to some extent admit, the contention implicit in many of his poems that human institutions are the cause of moral deficiencies and moral evils;

for if they do not create, they at least, so far as they are imperfect, tend to perpetuate them. The history of civilization is, in Shelley's view, a history of the unceasing struggle between the spirit of freedom and the spirit of tyranny, a struggle in which the nobler combatant has hitherto been the weaker. In his own day the cause of liberty has prospects brighter than ever before; yet, although the tyranny of force has fallen, that of fraud has succeeded, and this too must be striven with and overthrown. So long as his thoughts take a practical direction, Shelley is not only optimistic and full of hope for the future, but fully capable of estimating justly the existing obstacles to progress, and of suggesting sound and sensible reforms. But as a poet and theorist, his confidence in the perfectibility of human nature and of the conditions of its free development is more wavering and uncertain. In his poems the coming triumph upon earth of virtue and happiness is not prefigured as the consummation of a steadily advancing process; the reign of justice supervenes in an arbitrary and inexplicable manner upon ages of vice and misery. And we are left doubtful whether, after all, it is not upon another world than this that Shelley builds his hopes. Even in *Queen Mab* he speaks of Death as

> a gate of dreariness and gloom,
> That leads to azure isles and beaming skies,
> And happy regions of eternal hope—

and presumably it is behind these portals that the land of these glowing visions lies, which the Faery has conjured up before *Ianthe's* spirit. And in the

*Essay on Christianity* we read: 'Human life, with its unreal ills and transitory hopes, is as a dream which departs before the dawn, leaving no trace of its evanescent hues; all that it contains of pure or of divine visits the passive mind in some serenest dreams'; words which inevitably remind us of the oft-quoted lines in the *Sensitive Plant*,[1] and of the indistinct note in the last chorus of *Hellas*. Against these utterances, however, we may set the following passage in which Shelley speaks as a political reformer: 'We derive tranquillity and courage and grandeur of soul from contemplating an object which is because we will it, and may be, because we hope it and desire it, and must be if succeeding generations of the enlightened sincerely and earnestly desire it.' In this passage, written in 1819, we may perhaps discover a reconciliation of Shelley's practical optimism and theoretic idealism or transcendentalism.

The same conflict, and the same need for reconciliation, are seen in Shelley's conception of the manner in which reform and progress are to come. In the last-quoted passages he makes appeal to the general good-will of the community. But it would be dangerous to infer that Shelley regards the general good-will as the true and sole principle of advance. Both his idealism and his individualism forbid this. The conception of social progress most dear to Shelley is embodied in the image of Prometheus, bearing the unquenchable fire to earth, and in the symbol familiar to us in his other poems, an ardent and self-devoted

[1] For Love and Beauty and Delight
There is no change, &c.

spirit combating the united forces of tyranny and fraud, and departing, when once its work on earth is accomplished,

> Back to the burning fountain, whence it came.

To the same belief we must refer Shelley's admiration for the founders of the great religions, above all of Christianity.

For in religion, as in morals and politics, Shelley's antagonism is directed not against the living spirit and impulse, but against the concrete and transitory systems which profess to embody it. All particular religions, with the creeds and observances which they embrace, seem to him at best but the creation of the intellectual pride of man; at worst, they are the devices of fraud and tyranny. True religion, in the sense of a definite creed, can only exist as the genuine expression of the truths which stir our being most deeply. These truths, as Shelley conceives them, have been already indicated. The object of man's highest aspiration is not to his seeming a power actually present in the universe, working out its consummation through the energies of human thought and action, but a reality distinct and transcendent, a reality of which the solitary mountains are a type and symbol:—

> Power dwells alone in its serenity
> Remote, serene, and inaccessible.

This final reality is no common possession of the intellect, a thing to be discussed, defined, and labelled; it is a spirit dwelling apart, whose visitations are rare and only to the favoured. Such, at least, is the presentiment by which Shelley, in spite of his intense yearning for the triumph of universal goodness upon earth, is most persistently haunted.

If in his opinions on religion and metaphysics, on moral and social problems, Shelley has shown himself persistently idealistic, the same tendency may be expected of his conception of art in its essence and function. But Shelley's *Defence of Poetry*, from which the material for any theory of his aesthetic must chiefly be drawn, aims primarily at upholding the moral value of poetry against an insidious attack; and the essence of art and beauty is only investigated in relation to the main contentions. The two problems, however, are in fact inseparably connected, and it will perhaps not disturb the continuity of this preface if we consider the last and greatest of Shelley's prose writings in relation to its main thesis and purpose—as a vindication, that is, of the moral value of poetry and of all art.

The contest between art and morality, or rather between the artist and the would-be champion of public morals, has in all probability been waged since first man came to reflect upon his natural activities. It would doubtless appear (could the past but speak) that primitive attempts to embellish the weapons of war or of the chase were regarded with grave disapproval by the patriarch or public censor of the day, who would insist that the weapon was no better weapon, and its wielder no better warrior, for the time and labour thus expended upon it. The conflict once set on foot has maintained itself throughout historical ages, and there seems little likelihood that it will ever be resolved. In England it would have been waged with peculiar acrimony, were it not that the artist himself has often, perhaps too readily, disarmed criticism by ranging himself both in his theory and practice on the side of the moralist. And in that province of art whose borders are least definitely settled, the novel, our greatest writers have repeatedly and consciously proclaimed a moral purpose and mission. Moreover, the majority of those who have been provoked into a vindication of the arts which they practised, have failed, from lack of philosophical insight, to grasp the true nature of the problem : the real issues have been clouded or neglected, and any satisfactory solution precluded from the outset. In this respect Sidney and Shelley stand almost alone, by reason of their combination of speculative insight with productive power, as vindicators of the imaginative creation of man.

The real matter at issue between art and morality presents itself under many guises, but fundamentally it is always the same. Thus the question whether art ought to further the ends of morality, and the question whether it actually does so, can only be distinguished by an equivocal use of terms. It is illogical to demand of any human function an aim at variance with its nature, and thus all the questions properly resolve themselves into one, the problem, namely, whether true art is necessarily moral in its intention and influence. This question, it must be noted, is very different from the one which is often confounded with it, and regarded as the real issue at stake, the debate whether art in its various manifestations, good or bad, has actually promoted the good of mankind; whether men are better, worse, or neither better nor worse, for all that the artist has accomplished and left behind him.

With this central problem, indeed, it would be futile to deal without a clear conception of its meaning. It is one thing to assert that art will aid us in the performance of this or that particular duty; and quite another to hold that it enriches human consciousness, and has therefore a value for human life. This latter influence no man can reasonably deny to true art; should this fail to satisfy the moralist, he must, like Plato, banish the artists from his republic; for in demanding more of them he is, in effect, already banishing them. Similarly, the sense of this universal value of his work may reasonably be looked upon as the fundamental motive and stimulus to activity in the artist's mind; and the formula of art for art's sake be reasonably condemned, not merely on account of the emptiness of meaning which it reveals to a strict analysis, but because it is psychologically untrue. But it would not, therefore, be reasonable to expect of the artist a strictly didactic purpose, a purpose clearly defined by himself, and which animates and informs all his work.

But the majority of the champions of art, who have been artists or poets rather than logicians, have failed, as hinted above, to grasp the true significance of the primary demand, that art should promote morality, and have confined themselves to the attempt to prove, by reference to history and to experience, that the influence of art has been, on the whole, morally beneficial to mankind. Only in rare instances has the artist dared to claim a separate province and a distinct function for his activity. Even Sidney expresses himself with questionable clearness on this point. What, then, is Shelley's attitude to the matter?

The *Defence of Poetry* was written, as we have seen, to meet a particular attack, and Shelley's treatment of his subject is naturally influenced by the character of the attack which he is meeting. To Peacock's mock-historical account of the rise, climax, and degeneration of poetry, he opposes a brief *résumé* of its actual history. But the deeper aspects of the question could not fail to appeal to his philosophical habit of thought; and the historical account is prefaced by an apology based on metaphysical and psychological grounds. Neither in this part, however, nor elsewhere does he clearly enunciate the position, that morality in its narrower sense has no jurisdiction upon art. But here we must remember that morality, in this narrower sense, was, in Shelley's conception, not morality at all. Concrete systems of ethics interested him but little; the essence of morality lay, for him, not in the rigid application of a code of conduct, but in an ordered and harmonious condition of the soul. To this conception he will not provide a more definite content than by the opposition of the benevolent and malevolent feelings; these alone, he asserts, are essentially good and bad. This definition of morality enables him to posit, on metaphysical grounds, the identity of the beautiful and the good: for beauty, as that which it is the essential function of poetry to express, is the eternal order and rhythm of the universe; and a poem, in his definition, is 'the creation of actions according to the unchangeable forms of human nature, as existing in the mind of the creator, which is itself the image of all other minds'. And to the question whether the form of which he speaks is, in the work of art, the outward or sensuous form, or the idea or principle which it embodies, Shelley would apparently reply that the one involves and determines the other. Hence, all those in whom the idea is active are poets, whether they express it in verse, prose, or institution; for in all of these the inward harmony can be made manifest through the outward form. And hence it is that Shelley can interchange, with an apparent looseness of terminology, the terms beautiful, good, and true. 'To be a poet,' he says, 'is to apprehend the true and the beautiful, in a word, the good, which exists in the relation existing first between existence and perception, and secondly, between perception and expression.' The immutable order of things, as perceived, is truth; as expressed in art, it is beauty; as reflected in conduct, it is goodness.'[1]

In tracing the adumbrations of a metaphysic of beauty and these and similar passages scattered

[1] Cp. also (*Defence*): 'Poetry, in a restricted sense has a common source with all other forms of order and beauty'; and the 'poet participates in the eternal, the infinite, and the one.'

throughout Shelley's writings, it is impossible not to be impressed by their affinity to the Platonic theory of ideas. In fact, at the time when he wrote the *Defence*, Shelley was deeply immersed in the study of Plato, and had but lately translated his *Symposium*, that dialogue which exhibits Plato's idealism in its most characteristic, if not in its final form. The influence, indeed, of the doctrine is evident throughout the pages of the *Defence*. Yet this influence, profound as it undoubtedly was, is not to be traced entirely to the enthusiasm of the discipline in the revelation of new and unsuspected truth ; it is based upon a close affinity of temperament, indeed of actual spiritual history. The language in which in the *Symposium* Socrates asserts the transcendent reality of the Idea of Good, from which all things good and beautiful derive their essence, breathes the conviction which is engendered of direct personal experience. And we may perhaps, without being unduly literal in our interpretation, discover in Shelley's *Hymn to Intellectual Beauty* the record of an analogous experience, which gave the assurance of actual knowledge to what hitherto he had but imperfectly divined.[1] But whereas Plato, by the elaboration of his theory of ideas in the dialogues subsequent to the *Symposium*, gradually undermined his faith in the reality of the original vision, without attaining in compensation a solution of the crucial problem, Shelley confined himself to so much speculation upon his faith as gave it the

[1] See 'Plato's Vision of the Ideas', by W. Temple (*Mind*, Oct., 1908), where this analogy is pointed out.

semblance of philosophic doctrine, while not endangering it as a faith. Thus in the *Defence* and elsewhere he is content simply to assert the inherence of the absolute principle of order or beauty in particulars of sense, without attempting to define in scientific terms, as Plato had attempted it, the exact nature of the relationship. And it is upon the creed thus generally stated that Shelley bases his conviction that art and morality are ultimately one.

So much for the metaphysics of the question. As a problem of psychology, the relation of art and morals presents itself in two main aspects—in relation, first, to the creative impulse in the artist, and, secondly, to the nature of his influence upon others. To the first question Shelley's attitude is perhaps more apparently than actually clear or decisive. We may seek to elucidate it by reference both to his theory and to his practice, or rather to the principles by which, in his own belief, his practice was determined. His theory of the relation of the benevolent or altruistic, and the artistic motive, is briefly this. Love, or the principle of self-extension, lies at the root of both, and in either case its true organ is the imagination. 'It is because we enter into the meditations, designs, and destinies of something beyond ourselves,' declares the old man in *The Coliseum*, 'that the contemplation of the ruins of human power excites an elevating sense of awfulness and beauty. It is therefore that the ocean, the glacier, the cataract, the tempest, and the volcano, have each a spirit which animates the extremities of our frames with tingling joy. It is therefore that the singing of birds, and the motion of leaves, the sensation of the odorous earth beneath, and the freshness of the living wind around, is sweet. *And this is Love*'.[1] Thus the enjoyment, and still more the expression, of beauty implies the same identification of ourselves with what is without us and beyond, which lies at the root of the benevolent impulses. But to assert this is by no means to concede that the artist's motive, *qua* artist, is merely a specific instance of the wider motive of benevolence, or that the immediate purpose which guides his pen or pencil is the edification of his fellow men. The *result*, no doubt, of this self-diffusion which is achieved in art tends to human welfare, and of this the artist may be, nay, must be, ultimately conscious; but more than this Shelley, so far as we have followed him (and the above passage is a representative one), does not concede.[2]

But has he not, it may be asked, in stating the principles by which his own art is guided, come nearer to the moralist's position? At first sight it seems so. In the *Hymn to Intellectual Beauty*, the vow

> that I would dedicate my powers
> To thee and thine,

[1] The italics are mine. Cp. also (*Defence*): 'The great secret of morals is love: or a going out of our nature, and an identification of ourselves with the beautiful in thought, action, or person not our own.'

[2] Here again a comparison with Plato is invited. In the *Symposium*, love, or the principle of self-perpetuation, is represented as the root-impulse of all artistic, and also of all moral or social activity.

is tempered in its self-abandonment by its association with the hope

> that thou wouldst free
> This world from its dark slavery;

for the spells of Beauty are not of a kind to draw Shelley away from the world. On the contrary, they bind him 'to fear himself, and love all human kind'. And a sense of the close connexion in his own personal experience between the creative impulse and the impulse to help the world he loves, is evident in the language which he prefixes to many of his poems, as well as in the subject of the poems themselves. In *Laon and Cythna* his object, as stated expressly, is of enlisting the services of poetry 'in the cause of a liberal and comprehensive morality'. In the preface to *Prometheus Bound* he deprecates the erroneous assumption that he has dedicated his poetry 'solely to the direct enforcement of reform', a remark which implies a partial concession to that prejudice. In the same preface he defines more precisely his purpose hitherto as 'to familiarize the highly refined imagination of the more select classes of poetical readers with beautiful idealisms of moral excellence.' Assuming that Shelley did not deliberately depart in his practice from his own conception of the true poetic motive, we may reasonably infer from these passages that he regarded the philanthropic impulse as an essential element in the motives which determine the artist to create. But in drawing such a conclusion we must again be careful to remember that it is Shelley with whom we are dealing. His close identification of the beautiful with the good, on the one hand, and of the beautiful with the true, the ultimately real, on the other, might enable him to ascribe to the poet a function and motive essentially moral (in this widest sense of the term) without any detriment to his character as a poet. It was the delight in approximation to the real no less than to the good, and the desire to communicate that delight to others, that inspired Shelley to the portrayal of these 'beautiful idealisms of human excellence'. The moral content of a poem, in fact, was for him identical with the ideal (or real) content, that is, the expression of an ordered universe, in which the sum-total of human actions is expressive of law and reason, and of characters whose various elements are subordinated to a single rational principle. It is, in Shelley's judgement, by his power of exhibiting such a picture of the universe, combined with his 'bold neglect of a direct moral purpose', that Milton discloses in *Paradise Lost* the supremacy of his genius.

To attempt a critical examination of Shelley's views of poetical idealization would exceed the scope of this introduction; it is instructive, however, to notice that before he wrote *The Cenci* they seem to have been somewhat modified. In the preface to that poem he recognizes that the exigencies of dramatic effect may compel a more realistic treatment of life than had marked his former poems. Beatrice, the central figure of the drama, is neither conceived nor portrayed as an idealism of moral excellence. 'Had Beatrice thought in this manner' (i.e. more forgivingly), 'she would,' writes Shelley, 'have been wiser and better, but she would never have been a tragic character.' Moreover, in this drama Shelley handles 'a story eminently fearful and monstrous', a story in which all the actors are imperfect, and one is depraved above common measure. Shelley, in fact, is bringing poetry nearer to life;[1] and his account of his aim and method illustrates his sense of the changes that he has made. The ideal in this poem is not achieved through the presentation of patterns of moral excellence, but through 'the poetry which exists in these tempestuous sufferings and crimes'. In this latter phrase, though it would be unwise to press its meaning too far, Shelley seems to approach a more modern conception of the true nature of poetic idealization, as consisting in an interpretation of life which suppresses nothing essential, but which by emphasizing the significant traits and omitting the irrelevant in its subject-matter (be this, morally speaking, good or bad), attains a vividness of portraiture which actual experience never or rarely affords. Such idealization, be it observed, implies no sacrifice of moral effect. For 'the highest moral purpose aimed at in the highest species of the drama, is the teaching of the human heart, through its sympathies and antipathies, the knowledge of itself; in proportion

[1] Cp. letter to Leigh Hunt (1819):—'The drama which I now present to you is a sad reality. I lay aside the presumptuous attitude of the instructor, and am content to paint with such colours as my heart furnishes that which has been.' (Ingpen, *Letters of Shelley*, p. 690.)

to its possession of which knowledge, every human being is wise, just, sincere, tolerant, and kind.' Whether we call such a purpose moral or artistic is really indifferent; its true foundation is the desire of the individual, a desire deeply rooted in human nature, to share universally his deepest experiences.

And this leads us to that other aspect of the psychological problem. Granted that the true artist's motive is essentially moral, in what manner must that motive be realized in practice? Here Shelley speaks clearly enough. The narrow Puritanic conception that art must teach by conscious precept and illustration moves him repeatedly to eloquent protest: for it is responsible, in his opinion, for the whole misunderstanding between art and morality. 'The whole objection,' he writes in the *Defence*, 'of the immorality of poetry rests upon a misconception of the manner in which poetry acts to produce the moral improvement of man. . . . The great instrument of moral good is the imagination; and poetry administers to the effect by acting upon the cause. . . .' Poetry strengthens the faculty which is the organ of the moral nature of man, in the same manner as exercise strengthens a limb.' The vast, indeed exaggerated importance, which Shelley attaches to the imagination as a moral agent was touched upon in discussing his moral philosophy. Only by the relative compass of their imaginations, as we then saw, did Shelley distinguish the good man from the bad. Such a point of view is peculiarly opportune in a vindication of the moral value of poetry, and we cannot wonder that Shelley emphasizes it. For it enables him to contend that the poet, should he adopt any other method of communicating moral truths, would fail, not merely as a poet, but also as a moralist. The highest moral effect of poetry is in fact its effect as poetry. For 'those in whom the poetical faculty, though great, is less intense, as Euripides, Lucan, Tasso, Spenser, have frequently affected a moral aim, and the effect of their poetry is diminished in exact proportion to the degree in which they compel us to advert to this purpose'; but not, Shelley would add, in proportion to the degree in which they *effect* it.

On this question also Shelley is not least illuminating where he is giving account of his own poetic practice. In the preface to *Laon and Cythna*, in the composition of which poem he is more conscious than elsewhere of a moral purpose, he expressly states that he makes no attempt to enforce his views 'by methodical and systematic argument'. And in introducing *Prometheus Bound* to his readers he speaks more plainly still: 'Didactic poetry is my abhorrence; nothing can be equally well expressed in prose that is not tedious and supererogatory in verse.' He is 'aware that until the mind can love and admire and trust and hope and endure, reasoned principles of moral conduct are seeds cast upon the highway of life which the unconscious passenger tramples into dust, although they would bear the harvest of his happiness'. The truth embodied in these eloquent words is one which may well be recommended to-day to the advocate of moral instruction by direct precept, or of so-called 'aesthetic education'.

As to the manner in which imagination is to edify and elevate, Shelley's views would naturally vary with the change in his conception of poetic idealization, which we have already indicated. In the *Defence* he lays most stress on its power to familiarize us with types of noble humanity. It is because he 'embodied the ideal perfection of his age in human character' that Homer exercised so beneficent an influence; and Shelley's tribute to all the great poets of history rests in effect on the same basis. And it is in conformity with this conviction that he himself employs the imagination to present 'beautiful idealisms of moral excellence' to the world. In the preface to *The Cenci*, however, Shelley's conception of the poetic, and consequently of the moral, function of the imagination is modified. According to the passage already quoted, poetry must teach the human heart, through its sympathies *and* antipathies, the knowledge of itself. The moralist, no more than the poet, can afford to neglect or slur over the evidence of actual life. Such would seem to be Shelley's meaning in this preface; but we must remember that the attitude of the earlier prefaces, thus recalled in the *Defence*,[1] is the one most characteristic of his genius.

As a champion of the faculty of imagination, Shelley is almost inevitably involved in the *Defence* in a discussion of its claims relatively to those of reason.

[1] *recalled*; since the *Defence* was written some eighteen months after *The Cenci*.

Here again he wisely meets his antagonists on their own ground. Their plea being that, in respect of utility, reason must be allowed to be the superior faculty, Shelley points out that here, again, it is all a question of terms; only the narrowest conception of utility can deny the victory, even on this issue, to imagination. For 'the production and assurance of pleasure in this highest sense (the pleasure in poetical creations) is true utility. Those who produce and preserve this pleasure are poets or poetical philosophers'. The passage following, in which Shelley depreciates the work of the great rationalists and the services done by reason to mankind sounds strangely on the lips of Godwin's whilom disciple, yet both here and in the subsequent striking condemnation of the materialistic spirit of his own times we feel that it is the true Shelley who is speaking.

In the appeal to the testimony of history, which occupies about one half of the *Defence*, Shelley may seem to stand on less debatable ground than in the rest of his argument; but he makes, as a matter of fact, no attempt at a thoroughgoing historical inquiry, and the soundness of his views depends rather on the premises with which we are already familiar, than on the evidence of historical fact. With his intense conviction of the ultimate identity of man's ethical and imaginative activity, Shelley was pledged to discover the coincidence, which he claims in the *Defence*, between periods of moral and of artistic greatness. And it is perhaps only to a superficial view that the testimony of history seems on the whole to run counter to his assertion. It is true that periods which have produced the greatest masterpieces of art have been often periods, if not of moral decay, yet of moral disorder. But such disorder must necessarily attend each new awakening of the regenerative spirit in mankind, and will affect the province of art as well as that of conduct. The artist, however, having a less stubborn antagonism of prejudice and convention to contend with, realizes more speedily the new ideals and the new forms. Thus the upheaval in the sixteenth, and again in the eighteenth century was literary as well as social; but the masterpieces of Shakespeare and Goethe preceded the consummation of social and ethical reforms. Conversely, a social order once established may witness, before its own dissolution, the decline of literary or artistic movements to which it is organically related. With these reservations, it is not difficult to accept the dictum of Shelley, that 'the best poetry is contemporaneous with the noblest ages', or even to go further with him, and allow that poetry is a civilizer, as well as a mark of high civilization: that it 'contains within itself the seeds of its own and social renovation'. As Shelley subtly remarks of the corrupt poets of a corrupt age, 'it is not in what they have, but what they have not, that their imperfection consists.' So far as they are true poets, the influence even of these must be for good; while the nobleness of noble periods lives on in their masterpieces, which permanently preserve the energies and ideals not elsewhere accessible to a degenerate age.

The foregoing discussion of Shelley's theories of

art has limited itself to their general features ; a host of aesthetic problems, which are raised in the course of the *Defence*, have necessarily been left untouched. But enough has been said to mark, in this as in other branches of speculation, the idealizing quality of Shelley's genius, the aspiring temper which would elevate all reality to the level of a prefigured excellence. It may naturally be anticipated that a like spirit will pervade Shelley's actual literary and artistic criticism, his opinions, that is, upon particular works of art. Shelley's published writings are not rich in material of this kind, but the tenor of what exists will serve, on the whole, to confirm our expectations. Nothing in this connexion is more pertinent than Shelley's intense and often avowed admiration for the genius of ancient Greece ; an admiration not confined to their work as artists, but embracing the whole of their civilization. 'But for those changes,' he writes in the year 1818, 'that conducted Athens to its ruin, to what an eminence might not humanity have arrived.' [1] It is, however, for the art of Athens, above all for its sculpture, that Shelley's warmest sympathy is reserved. And it is not difficult to see where the roots of this sympathy lie. 'I now understand' (to quote once more Shelley's own words) 'why the Greeks were such great artists, and above all I can account, as it seems to me, for the harmony, the perfection, the uniform excellence of their works of art. They lived in a perpetual commerce with Nature, and nourished themselves upon the spirit of her forms.' And elsewhere he calls their sculptures

[1] And at the end of 1819 he writes in the same strain.

'models of ideal truth and beauty'. Such language seems peculiarly appropriate; for by the directness and simplicity of his aim, his unerring vision and unrivalled powers of execution, the Greek sculptor succeeded, more completely than the artist of any period, in subjugating his material to his purpose and in producing the effect of a nature dematerialized, of sense wholly translated into thought. The fact that this end is attained at the expense of valuable qualities—of character, individuality, life—does not affect Shelley as it affects the modern consciousness in general. Thus even Goethe could write that 'the generic conception (in Greek art) leaves us cold'. Such a judgement would not have commended itself to Shelley, whose descriptions of the sculptures visited by him in Italy betoken a warm and exalted enthusiasm. Yet his consciousness of the very spirit of beauty imprisoned, as it were, in these creations, does not overcloud his insight into their meaning, or tempt him to the fallacious conclusion that such beauty was the object of the artist's deliberate endeavour. Everywhere he recognizes as the purpose of Greek statuary the expression of certain phases of conscious life, and it is its perfect achievement of this purpose that wins his critical approval. But the range of this expression is limited. 'The coarser and more violent effects of comic feeling,' writes Shelley, 'cannot be seized by this art. Tenderness, sensibility, enthusiasm, terror, poetic inspiration, the profound, the beautiful, Yes.' And in the lofty position assigned by him to an art thus restricted, we have new evidence of the limitations of Shelley's own philosophy of art; of his tendency, that is, to confine beauty in art to the expression of an ideal harmony and order. Thus, of the face of the central figure in the *Niobe* he writes that 'it resembles the careless majesty which Nature stamps upon the rare masterpieces of her creation, harmonizing them as it were from the harmony of the spirit within. Yet all this not only consists with, but is the cause of the subtlest delicacy of clear and tender beauty'. Here Shelley speaks, it is true, of the order existing in the human soul; but this itself is to him but a reflection of the universal harmony.

It might well be expected that Shelley's enthusiasm for the classical ideal would narrow his sympathies and obscure his judgement in the presence of modern works of art. And his criticisms of the sculpture and paintings of Italy to some extent justify such an anticipation. Thus his distaste for Michel-Angelo's work is evidently founded upon its violation of those canons of beauty, of which Greek sculpture had afforded the perfect exemplar. 'He has not only (so Shelley judges) no temperance, no modesty, no feelings for the just boundaries of art (and in these respects an admirable genius may err), but he has no sense of beauty, and to want this is to want the sense of the creative power of mind. What a thing his Moses is, how distorted from all that is natural and majestic!' And, again, 'He seems to me to have no sense of moral dignity and loveliness.' To this depreciation of Michel-Angelo corresponds an undue exaltation of Guido Reni, and, it may perhaps be added, of Raphael.

The paintings of these artists are (with the strange addition of Salvator Rosa) the only things which in Shelley's eyes 'sustain the comparison with antiquity.' Of Raphael's St. Cecilia he writes, 'It is of the inspired and ideal kind, and seems to have been executed in a similar state of feeling to that in which the ancients produced those perfect specimens of poetry and sculpture, which are the baffling models of succeeding generations.' Of Raphael such praise may seem not extravagant to the critic of to-day, but he will doubtless find it strange that the ideal loveliness, such as it is, of Guido's work should have blinded Shelley to its lack of real sincerity and strength.

Remembering, however, how entirely Shelley was thrown upon his own untutored judgement in matters of art, we must acknowledge that his intuitive perception of the great qualities of Greek art, though occasionally it may have misled him, was on the whole far more of a stimulus than an impediment to the growth of his critical faculty. In literature he was somewhat differently situated. His taste had been to a great extent formed upon other models than those of Greece; it was in consequence more catholic, and acknowledged wider canons. Yet even in literature, Shelley's fidelity to his ideal of an outward harmony, embodying the inward, is evident. Hence it is that in this province of art also he inclines to award the palm to the ancient Greeks. 'Perhaps Shakespeare,' he writes, 'from the variety and comprehension of his genius, is to be considered the greatest individual mind of which we have specimens remaining; but as a poet Homer must be acknowledged to exceed Shakespeare in the truth and harmony, the sustained grandeur, and satisfying completeness of his images, their exact fitness to the illustration and to that to which they belong.'[1] And Homer's subject, as we learn elsewhere, was 'the ideal perfection of his age' embodied in human character. Similarly Milton's greatness consists in his arrangement 'of the elements of human nature according to the laws of that principle by which a series of actions and of the external universe is calculated to excite the sympathy of succeeding generations of mankind.' And Milton himself falls short, according to Shelley, of the ideal in his portrait of Satan, who as a poetic figure is inferior to Prometheus by reason of 'the inward disorder arising from envy, ambition, revenge, and desire for personal aggrandizement'. The same attitude may perhaps be traced in his intense admiration of Dante, 'who excelled all poets except Shakespeare in ideal beauty'; of Calderon, who seemed to him 'a kind of Shakespeare'; or of Boccaccio, who 'possessed a deep sense of the fair ideal of human life'.

It would, however, be unwise to press too far this aspect of Shelley's literary sympathies, which are too wide and varied to admit of being referred to a single dominant criterion. What we have chiefly to remember of him as a critic is, first, his almost unfailing

[1] More significant is the passage following this (in the *Manners of the Athenians*, p. 34), 'omitting the comparison of individual minds, which can afford no general inference, how superior was the spirit and system of their poetry to that of any other period!'

sense for great and enduring qualities in works of art, wheresoever he meets them, and secondly, his entire freedom from personal or national bias and prejudice. In this respect he compares favourably with some of his contemporaries. It was no small thing, at a time when Goethe was known in England chiefly as the author of *Werter*, and when even such critics as Wordsworth, Coleridge, and Lamb had little to say in his favour, that Shelley should both have admired him sincerely and openly acknowledged his admiration. The same generosity of judgement he displayed towards the living writers of his own country. While he disliked Wordsworth's politics, he recognized the greatness of his poetry, and did not deny its influence upon his own. Coleridge was to him a figure of central interest among contemporary writers. For Byron's powers, while he lamented their aberrations, he entertained the profoundest admiration : nor did his aversion from the 'canons of taste' which disfigured Keats's earlier composition blind him to the evidence of transcendent genius in the *Hyperion*. And if in reviewing the work of Godwin, Peacock, or Hogg, he seems at times unduly laudatory, we must perhaps remember that personal feeling, while it could in no case become with Shelley a motive for detraction, might unconsciously sway him to a too favourable opinion, though never to an opinion wholly unsupported by his critical judgement. Finally, in his harsh condemnation of French literature, it is not unlikely that Shelley has the models of the classical period in mind, and contrasts the writers who attained formal correctness at the expense of inspiration, with one who seemed, in spite of artistic shortcomings, essentially a poet at heart.[1]

On the whole, indeed, Shelley's literary judgements are characterized by a breadth and tolerance widely at variance with the popular conception of his genius. This quality reveals itself in all his more mature and independent speculation. 'It is necessary,' he writes in the *Essay on Christianity*, 'that universal benevolence should supersede the regulations of precedent and prescription, before these regulations can safely be abolished.' And in the same essay (speaking of the progress of reason), 'It is useful to a certain extent that they (mankind) should not consider these institutions which they have been habituated to reverence as opposing an obstacle to its advance.' And again, writing of the present state of England, he declares that 'the great thing to do is to hold the balance between popular impatience and tyrannical obstinacy'. Such language as this is far indeed from the temper of the mere visionary or iconoclast. And the same desire for scientific impartiality marks his discussion of the various fundamental problems embraced in these pages—of life, of punishment by death, of a future state. Even when, as in the essay on love, Shelley is speaking directly from an intimate personal experience, he yet seeks to bring that experience into accord with elementary facts of nature and human consciousness. And if, as it has been the tenor of this introduction to indicate, the task of reconciling the

[1] i. e. Rousseau. 'French literature, which the great name of Rousseau alone redeems'; and again, 'Rousseau is a poet.'

ideal and real lay outside the scope of Shelley's genius, yet it is no small thing that he saw the necessity of its accomplishment. By embodying and illustrating this conviction Shelley's prose writings are a necessary complement to his poems; and for this alone they would deserve the attention of all to whom his message is of import. As for their intrinsic literary quality, their inalienable charm of style, the appreciation of this can only be endangered by inadequate eulogy or analysis.

J. SHAWCROSS.

## NOTE

The text of the present volume is based upon the following editions of Shelley's writings :—*Essays and Letters of Shelley*, edited by Mrs. Shelley, 1852; *Shelley Memorials*, edited by Lady Shelley, 1859; and *Shelley's Prose Works*, in four volumes, edited by H. Buxton Forman, 1880. The fragment entitled *A Fable* is taken from *Relics of Shelley*, edited by Richard Garnett, 1862, where the original version in Italian and the editor's translation (which alone is here reproduced) will be found together.

To Mr. Buxton Forman my best thanks are due for allowing me to make use of his valuable edition, which I have also followed in assigning dates to the various compositions.

J. S.

# CONTENTS

# REMARKS ON *MANDEVILLE* AND MR. GODWIN

The author of *Mandeville* is one of the most illustrious examples of intellectual power of the present age. He has exhibited that variety and universality of talent which distinguishes him who is destined to inherit lasting renown, from the possessors of temporary celebrity. If his claims were to be measured solely by the accuracy of his researches into ethical and political science, still it would be difficult to name a contemporary competitor. Let us make a deduction of all those parts of his moral system which are liable to any possible controversy, and consider simply those which only to allege is to establish, and which belong to that most important class of truths which he that announces to mankind seems less to teach than to recall.

*Political Justice* is the first moral system explicitly founded upon the doctrine of the negativeness of rights and the positiveness of duties,—an obscure feeling of which has been the basis of all the political liberty and private virtue in the world. But he is also the author of *Caleb Williams*; and if we had no record of a mind, but simply some fragment containing the conception of the character of Falkland, doubtless we should say, 'This is an extraordinary mind, and undoubtedly was capable of the very sublimest enterprises of thought.'

St. Leon and Fleetwood are moulded with somewhat inferior distinctness, in the same character of a union of delicacy and power. The Essay on Sepulchres has all the solemnity and depth of passion which belong to a mind that sympathizes, as one man with his friend in the interest of future ages, in the concerns of the vanished generations of mankind.

It may be said with truth, that Godwin has been treated unjustly by those of his countrymen, upon whose favour temporary distinction depends. If he had devoted his high accomplishments to flatter the selfishness of the rich, or enforced those doctrines on which the powerful depend for power, they would, no doubt, have rewarded him with their countenance, and he might have been more fortunate in that sunshine than Mr. Malthus or Dr. Paley. But the difference would have been as wide as that which must for ever divide notoriety from fame. Godwin has been to the present age in moral philosophy what Wordsworth is in poetry. The personal interest of the latter would probably have suffered from his pursuit of the true principles of taste in poetry, as much as all that is temporary in the fame of Godwin has suffered from his daring to announce the true foundations of minds, if servility, and dependence, and superstition, had not been too easily reconcilable with his species of dissent from the opinions of the great and the prevailing. It is singular that the other nations of Europe should have anticipated, in this respect, the judgement of posterity; and that the name of Godwin and that of his late illustrious and admirable wife, should be pronounced, even by those who know but little of English literature, with reverence and admiration; and that the writings of Mary Wollstonecraft should have been translated, and universally read, in France and Germany, long after the bigotry of faction has stifled them in our own country.

*Mandeville* is Godwin's last production. In interest it is perhaps inferior to *Caleb Williams*. There is no character like Falkland, whom the author, with that sublime casuistry which is the parent of toleration and forbearance, persuades us personally to love, whilst his actions must for ever remain the theme of our astonishment and abhorrence. Mandeville challenges our compassion, and no more. His errors arise from an immutable necessity of internal nature, and from much constitutional antipathy and suspicion, which soon spring up into hatred and contempt, and barren misanthropy, which, as it has no root in genius or virtue, produces no fruit uncongenial with the soil wherein it grew. Those of Falkland sprang from a high, though perverted conception of human nature, from a powerful sympathy with his species, and from a temper which led him to believe that the very reputation of excellence should walk among mankind unquestioned and unassailed. So far as it was a defect to link the interest of the tale with anything inferior to Falkland, so is Mandeville defective. But the varieties of human character, the depth and complexity of human motive,—those sources of the union of strength and weakness—those powerful sources of pleading for universal kindness and toleration,—are just subjects for illustration and development in a work of fiction; as such, *Mandeville* yields in interest and importance to none of the productions of the author. The events of the tale flow like the stream of fate, regular and irresistible, growing at once darker and swifter in their progress: there is no surprise, no shock: we are prepared for the worst from the very opening of the scene, though we wonder whence the author drew the shadows which render the moral darkness, every instant more fearful, at last so appalling and so complete. The interest is awfully deep and rapid. To struggle with it, would be the gossamer attempting to bear up against the tempest.

In this respect it is more powerful than *Caleb Williams*; the interest of *Caleb Williams* being as rapid, but not so profound, as that of *Mandeville.* It is a wind that tears up the deepest waters of the ocean of mind.

The language is more rich and various, and the expressions more eloquently sweet, without losing that energy and distinctness which characterize *Political Justice* and *Caleb Williams.* The moral speculations have a strength, and consistency, and boldness, which has been less clearly aimed at in his other works of fiction. The pleadings of Henrietta to Mandeville, after his recovery from madness, in favour of virtue and of benevolent energy, compose, in every respect, the most perfect and beautiful piece of writing of modern times. It is the genuine doctrine of *Political Justice*, presented in one perspicacious and impressive river, and clothed in such enchanting melody of language, as seems, not less than the writings of Plato, to realize those lines of Milton :

> How charming is divine philosophy—
> Not harsh and crabbed—
> But musical as is Apollo's lute !

Clifford's talk, too, about wealth, has a beautiful, and readily to be disentangled intermixture of truth and error. Clifford is a person, who, without those characteristics which usually constitute the sublime, is sublime from the mere excess of loveliness and innocence. Henrietta's first appearance to Mandeville, at Mandeville House, is an occurrence resplendent with the sunrise of life ; it recalls to the memory many a vision—or perhaps but one—which the delusive exhalations of unbaffled hope have invested with a rose-like lustre as of morning, yet unlike morning—a light which, once extinguished, never can return. Henrietta seems at first to be all that a susceptible heart imagines in the object of its earliest passion. We scarcely can see her, she is so beautiful. There is a mist of dazzling loveliness which encircles her, and shuts out from the sight all that is mortal in her transcendent charms. But the veil is gradually undrawn, and she 'fades into the light of common day'. Her actions, and even her sentiments, do not correspond to the elevation of her speculative opinions, and the fearless sincerity which should be the accompaniment of truth and virtue. But she has a divided affection, and she is faithful there only where infidelity would have been self-sacrifice. Could the spotless Henrietta have subjected her love to Clifford, to the vain and insulting accident of wealth and reputation, and the babbling of a miserable old woman, and yet have proceeded unshrinking to her nuptial feast from the expostulations of Mandeville's impassioned and pathetic madness? It might be well in the author to show the foundations of human hope thus overthrown, for his picture might otherwise have been illumined with one gleam of light. It was his skill to enforce the moral, 'that all things are vanity,' and 'that the house of mourning is better than the house of feasting'; and we are indebted to those who make us feel the instability of our nature, that we may lay the knowledge (which is its foundation) deep, and make the affections (which are its cement) strong. But one regrets that Henrietta,—who soared far beyond her contemporaries in her opinions, who was so beautiful that she seemed a spirit among mankind,—should act and feel no otherwise than the least exalted of her sex; and still more, that the author, capable of conceiving something so admirable and lovely, should have been withheld, by the tenour of the fiction which he chose, from executing it in its full extent. It almost seems in the original conception of the character of Henrietta, that something was imagined too vast and too uncommon to be realized ; and the feeling weighs like disappointment on the mind. But these objections, considered with reference to the close of the story, are extrinsical.

The reader's mind is hurried on as he approaches the end with breathless and accelerated impulse. The noun *smorfia* comes at last, and touches some nerve which jars the inmost soul, and grates, as it were, along the blood ; and we can scarcely believe that that grin which must accompany Mandeville to his grave, is not stamped upon our own visage.

1816.

## ON *FRANKENSTEIN*

THE novel of *Frankenstein ; or, The Modern Prometheus*, is undoubtedly, as a mere story, one of the most original and complete productions of the day. We debate with ourselves in wonder, as we read it, what could have been the series of thoughts—what could have been the peculiar experiences that awakened them—which conduced, in the author's mind, to the astonishing combinations of motives and incidents, and the startling catastrophe, which compose this tale. There are, perhaps, some points of subordinate importance, which prove that it is the author's first attempt. But in this judgement, which requires a very nice discrimination, we may be mistaken ; for it is conducted throughout with a firm and steady hand. The interest gradually accumulates and advances towards the conclusion with the accelerated rapidity of a rock rolled down a mountain. We are led breathless with suspense and sympathy, and the heaping up of incident on incident, and the working of passion out of passion. We cry 'hold,

hold! enough!'—but there is yet something to come; and, like the victim whose history it relates, we think we can bear no more, and yet more is to be borne. Pelion is heaped on Ossa, and Ossa on Olympus. We climb Alp after Alp, until the horizon is seen blank, vacant, and limitless; and the head turns giddy, and the ground seems to fail under our feet.

This novel rests its claim on being a source of powerful and profound emotion. The elementary feelings of the human mind are exposed to view; and those who are accustomed to reason deeply on their origin and tendency will, perhaps, be the only persons who can sympathize, to the full extent, in the interest of the actions which are their result. But, founded on nature as they are, there is perhaps no reader, who can endure anything beside a new love-story, who will not feel a responsive string touched in his inmost soul. The sentiments are so affectionate and so innocent—the characters of the subordinate agents in this strange drama are clothed in the light of such a mild and gentle mind—the pictures of domestic manners are of the most simple and attaching character: the father's is irresistible and deep. Nor are the crimes and malevolence of the single Being, though indeed withering and tremendous, the offspring of any unaccountable propensity to evil, but flow irresistibly from certain causes fully adequate to their production. They are the children, as it were, of Necessity and Human Nature. In this the direct moral of the book consists; and it is perhaps the most important, and of the most universal application, of any moral that can be enforced by example. Treat a person ill, and he will become wicked. Requite affection with scorn;—let one being be selected, for whatever cause, as the refuse of his kind—divide him, a social being, from society, and you impose upon him the irresistible obligations—malevolence and selfishness. It is thus that, too often in society, those who are best qualified to be its benefactors and its ornaments, are branded by some accident with scorn, and changed, by neglect and solitude of heart, into a scourge and a curse.

The Being in *Frankenstein* is, no doubt, a tremendous creature. It was impossible that he should not have received among men that treatment which led to the consequences of his being a social nature. He was an abortion and an anomaly ; and though his mind was such as its first impressions framed it, affectionate and full of moral sensibility, yet the circumstances of his existence are so monstrous and uncommon, that, when the consequences of them became developed in action, his original goodness was gradually turned into inextinguishable misanthropy and revenge. The scene between the Being and the blind De Lacey in the cottage, is one of the most profound and extraordinary instances of pathos that we ever recollect. It is impossible to read this dialogue,—and indeed many others of a somewhat similar character,—without feeling the heart suspend its pulsations with wonder, and the ‘ tears stream down the cheeks ’. The encounter and argument between Frankenstein and the Being on the sea of ice, almost approaches, in effect, to the expostulation of Caleb Williams with Falkland. It reminds us, indeed, somewhat of the style and character of that admirable writer, to whom the author has dedicated his work, and whose productions he seems to have studied.

There is only one instance, however, in which we detect the least approach to imitation ; and that is the conduct of the incident of Frankenstein's landing in Ireland. The general character of the tale, indeed, resembles nothing that ever preceded it. After the death of Elizabeth, the story, like a stream which grows at once more rapid and profound as it proceeds, assumes an irresistible solemnity, and the magnificent energy and swiftness of a tempest.

The churchyard scene, in which Frankenstein visits the tombs of his family, his quitting Geneva, and his journey through Tartary to the shores of the Frozen Ocean, resemble at once the terrible reanimation of a corpse and the supernatural career of a spirit. The scene in the cabin of Walton's ship—the more than mortal enthusiasm and grandeur of the Being's speech over the dead body of his victim—is an exhibition of intellectual and imaginative power, which we think the reader will acknowledge has seldom been surpassed.

1817.

# PRINCE ALEXY HAIMATOFF

[*Memoirs of Prince Alexy Haimatoff*. Translated from the original Latin MSS. under the immediate inspection of the Prince. By John Brown, Esq. Pp. 236, 12mo. Hookham, 1814.]

Is the suffrage of mankind the legitimate criterion of intellectual energy ? Are complaints of the aspirants to literary fame to be considered as the honourable disappointment of neglected genius, or the sickly impatience of a dreamer miserably self deceived ? the most illustrious ornaments of the annals of the human race have been stigmatized by the contempt and abhorrence of entire communities of man ; but this injustice arose out of some temporary superstition, some partial interest, some national doctrine : a glorious redemption awaited their remembrance. There is indeed, nothing so remarkable in the contempt of the ignorant for the enlightened : the vulgar pride of folly delights to triumph upon mind. This is an intelligible process : the infamy or ingloriousness that can be thus explained detracts nothing from the beauty of virtue or the sublimity of genius. But what does utter obscurity express ? if the public do not advert even in censure to a performance, has that performance already received its condemnation ?

The result of this controversy is important to the ingenuous critic. His labours are indeed miserably worthless if their objects may invariably be attained before their application. He should know the limits of his prerogative. He should not be ignorant, whether it is his duty to promulgate the decisions of others, or to cultivate his taste and judgement, that he may be enabled to render a reason for his own.

Circumstances the least connected with intellectual nature have contributed, for a certain period, to retain in obscurity the most memorable specimens of human genius. The author refrains perhaps from introducing his production to the world with all the pomp of empirical bibliopolism. A sudden tide in the affairs of men may make the neglect or contradiction of some insignificant doctrine a badge of obscurity and discredit : those even who are exempt from the action of these absurd predilections are necessarily in an indirect manner affected by their influence. It is perhaps the product of an imagination daring and undisciplined : the majority of readers ignorant and disdaining toleration refuse to pardon a neglect of common rules ; their canons of criticism are carelessly infringed, it is less religious than a charity sermon, less methodical and cold than a French tragedy, where all the unities are preserved : no excellencies, where prudish cant and dull regularity are absent, can preserve it from the contempt and abhorrence of the multitude. It is evidently not difficult to imagine an instance in which the most elevated genius shall be recompensed with neglect. Mediocrity alone seems unvaryingly to escape rebuke and obloquy, it accommodates its attempts to the spirit of the age which has produced it, and adopts with mimic effrontery the cant of the day and hour for which alone it lives.

We think that *The Memoirs of Prince Alexy Haimatoff* deserves to be regarded as an example of the fact by the frequency of which criticism is vindicated from the imputation of futility and impertinence. We do not hesitate to consider this fiction as the product of a bold and original mind. We hardly remember ever to have seen surpassed the subtle delicacy of imagination, by which the manifest distinctions of character and form are seized and pictured in colours that almost make nature more beautiful than herself. The vulgar observe no resemblances or discrepancies, but such as are gross and glaring. The science of mind to which history, poetry, biography serve as the materials, consists in the discernment of shades and distinctions where the unenlightened discover nothing but a shapeless and unmeaning mass. The faculty for this discernment distinguishes genius from dullness. There are passages in the production before us which afford instances of just and rapid intuition belonging only to intelligences that possess this faculty in no ordinary degree. As a composition the book is far from faultless. Its abruptness and angularities do not appear to have received the slightest polish or correction. The author has written with fervour, but has disdained to revise at leisure. These errors are the errors of youth and genius and the fervid impatience of sensibilities impetuously disburthening their fullness. The author is proudly negligent of connecting the incidents of his tale. It appears more like the recorded day dream of a poet, not unvisited by the sublimest and most lovely visions, than the tissue of a romance skilfully interwoven for the purpose of maintaining the interest of the reader, and conducting his sympathies by dramatic gradations to the *dénouement.* It is, what it professes to be, a memoir, not a novel. Yet its claims to the former appellation are established, only by the impatience and inexperience of the author, who, possessing in an eminent degree, the higher qualifications of a novelist, we had almost said a poet, has neglected the number by which that success would probably have been secured, which, in this instance, merits of a far nobler stamp have unfortunately failed to acquire. Prince Alexy is by no means an unnatural, although no common character. We think we can discern his counterpart in Alfieri's delineation of himself. The same propensities, the same ardent devotion to his purposes, the same chivalric and unproductive attachment to unbounded liberty, characterizes both. We are inclined to doubt whether the author has not attributed to his hero the doctrines of universal philanthropy in a spirit of profound and almost unsearchable irony: at least he appears biased by no peculiar principles, and it were perhaps an insoluble inquiry whether any, and if any, what moral truth he designed to illustrate by his tale. Bruhle, the tutor of Alexy, is a character delineated with consummate skill; the power of intelligence and virtue over external deficiencies is forcibly exemplified. The calmness, patience and magnanimity of this singular man, are truly rare and admirable: his disinterestedness, his equanimity, his irresistible gentleness, form a finished and delightful portrait. But we cannot regard his commendation to his pupil to indulge in promiscuous concubinage without horror and detestation. The author appears to deem the loveless intercourse of brutal appetite a venial offence against delicacy and virtue! he asserts that a transient connexion with a cultivated female may contribute to form the heart without essentially vitiating the sensibilities. It is our duty to protest against so pernicious and disgusting an opinion. No man can rise pure from the poisonous embraces of a prostitute, or sinless from the desolated hopes of a confiding heart. Whatever may be the claims of chastity, whatever the advantages of simple and pure affections, these ties, these benefits, are of equal obligation to either sex. Domestic relations depend for their integrity upon a complete reciprocity of duties. But the author himself has in the adventure of the Sultana, Debesh-Sheptuti, afforded a most impressive and tremendous allegory of the cold-blooded and malignant selfishness of sensuality.

We are incapacitated by the unconnected and vague narrative from forming an analysis of the incidents: they would consist indeed, simply of a catalogue of events, and which, divested of the aerial tinge of genius, might appear trivial and common. We shall content ourselves, therefore, with selecting some passages calculated to exemplify the peculiar powers of the author. The following description of the simple and interesting Rosalie is in the highest style of delineation :—

Her hair was unusually black, she truly had raven locks, the same glossiness, the same varying shade, the same mixture of purple and sable for which the plumage of the raven is remarkable, were found in the long elastic tresses depending from her head and covering her shoulders. Her complexion was dark and clear; the colours which composed the brown that dyed her smooth skin, were so well mixed, that not one blot, not one varied tinge, injured its brightness, and when the blush of animation or of modesty flushed her cheek, the tint was so rare, that could a painter have dipped his pencil in it, that single shade would have rendered him immortal. The bone above her eye was sharp, and beautifully curved; much as I have admired the wonderful properties of curves, I am convinced that their most stupendous properties collected would fall far short of that magic line. The eyebrow was pencilled with extreme nicety; in the centre it consisted of the deepest shade of black, at the edges it was hardly perceptible, and no man could have been hardy enough to have attempted to define the precise spot at which it ceased: in short the velvet drapery of the eyebrow was only to be rivalled by the purple of the long black eyelashes that terminated the ample curtain. Rosalie's eyes were large and full; they appeared at a distance uniformly dark, but upon close inspection the innumerable strokes of various hues of infinite fineness and endless variety, drawn in concentric circles behind the pellucid crystal, filled the mind with wonder and admiration, and could only be the work of infinite power directed by infinite wisdom.

Alexy's union with Aür-Ahebeh the Circassian slave is marked by circumstances of deep pathos, and the sweetest tenderness of sentiment. The description of his misery and madness at her death deserves to be remarked as affording evidence of an imagination vast, profound and full of energy.

Alexy, who gained the friendship, perhaps the love of the native Rosalie: the handsome Haimatoff, the philosophic Haimatoff, the haughty Haimatoff, Haimatoff the gay, the witty, the accomplished, the bold hunter, the friend of liberty, the chivalric lover of all that is feminine, the hero, the enthusiast: see him now, that is he, mark him! he appears in the shades of evening, he stalks as a spectre, he has just risen from the damps of the charnel-house; see, the dews still hang on his forehead. He will vanish at cock-crowing, he never heard the song of the lark, nor the busy hum of men; the sun's rays never warmed him, the pale moonbeam alone shows his unearthly figure, which is fanned by the wing of the owl, which scarce obstructs the slow flight of the droning beetle, or of the drowsy bat. Mark him! he stops, his lean arms are crossed on his bosom; he is bowed to the earth, his sunken eye gazes from its deep cavity on vacuity, as the toad skulking in the corner of a sepulchre, peeps with malignity through the circumambient gloom. His cheek is hollow; the glowing tints of his complexion, which once resembled the autumnal sunbeam on the autumnal beech, are gone, the cadaverous yellow, the livid hue, have usurped their place, the sable honours of his head have perished, they once waved in the wind like the jetty pinions of the raven, the skull is only covered by the shrivelled skin, which the rook views wistfully, and calls to her young ones. His gaunt bones start from his wrinkled garments, his voice is deep, hollow, sepulchral; it is the voice which wakes the dead, he has long held converse with the departed. He attempts to walk he knows not whither, his legs totter under him, he falls, the boys hoot him, the dogs bark at him, he hears them not, he sees them not.—Rest there, Alexy, it beseemeth thee, thy bed is the grave, thy bride is the worm, yet once thou stoodest erect, thy cheek was flushed with joyful ardour, thy eye blazing told what thy head conceived, what thy heart felt, thy limbs were vigour and activity, thy bosom expanded with pride, ambition, and desire, every nerve thrilled to feel, every muscle swelled to execute.

Haimatoff, the blight has tainted thee, thou ample roomy web of life, whereon were traced the gaudy characters, the gay embroidery of pleasure, how has the moth battened on thee; Haimatoff, how has the devouring flame scorched the plains, once yellow with the harvest! the simoon, the parching breath of the desert, has swept over the laughing plains, the carpet of verdure rolled away at its approach, and has bared amid desolation. Thou stricken deer, thy leather coat, thy dappled hide hangs loose upon thee, it was a deadly arrow, how has it wasted thee, thou scathed oak, how has the red lightning drank thy sap: Haimatoff, Haimatoff, eat thy soul with vexation. Let the immeasurable ocean roll between thee and pride: you must not dwell together, p. 129.

The episode of Viola is affecting, natural, and beautiful. We do not ever remember to have seen the unforgiving fastidiousness of family honour more awfully illustrated. After the death of her lover, Viola still expects that he will esteem [1], still cherishes the delusion that he is not lost to her for ever.

She used frequently to go to the window to look for him, or walk in the Park to meet him, but without the least impatience, at his delay. She learnt a new tune, or a new song to amuse him, she stood behind the door to startle him as he entered, or disguised herself to surprise him.

The character of Mary, deserves, we think, to be considered as the only complete failure in the book. Every other female whom the author has attempted to describe is designated by an individuality peculiarly marked and true. They constitute finished portraits of whatever is eminently simple, graceful, gentle, or

[1] ? misprint for 'return' [Ed.].

disgustingly atrocious and vile. Mary alone is the miserable parasite of fashion, the tame slave of drivelling and drunken folly, the cold-hearted coquette, the lying and meretricious prude. The means employed to gain this worthless prize corresponds exactly with its worthlessness. Sir Fulke Hildebrand is a strenuous Tory, Alexy, on his arrival in England professes himself inclined to the principles of the Whig party, finding that the Baronet had sworn that his daughter should never marry a Whig, he sacrifices his principles and with inconceivable effrontery thus palliates his apostasy and falsehood.

> The prejudices of the Baronet were strong in proportion as they were irrational. I resolved rather to humour than to thwart them. I contrived to be invited to dine in company with him; I always proposed the health of the minister, I introduced politics and defended the Tory party in long speeches, I attended clubs and public dinners of that interest. I do not know whether this conduct was justifiable; it may certainly be excused when the circumstances of my case are duly considered. I would tear myself in pieces if I suspected that I could be guilty of the slightest falsehood or prevarication; (see Lord Chesterfield's Letters for the courtier-like distinction between simulation and dissimulation) but there was nothing of that sort here. I was of no party, consequently, I could not be accused of deserting any one. I did not defend the injustice of any body of men, I did not detract from the merits of any virtuous character. I praised what was laudable in the Tory party, and blamed what was reprehensible in the Whigs: I was silent with regard to whatever was culpable in the former or praiseworthy in the latter. The stratagem was innocent which injured no one, and which promoted the happiness of two individuals, especially of the most amiable woman the world ever knew.

An instance of more deplorable perversity of the human understanding we do not recollect ever to have witnessed. It almost persuades us to believe that scepticism or indifference concerning certain sacred truths may occasionally produce a subtlety of sophism, by which the conscience of the criminal may be bribed to overlook his crime.

Towards the conclusion of this strange and powerful performance it must be confessed that *aliquando bonus dormitat Homerus.* The adventure of the Eleutheri, although the sketch of a profounder project, is introduced and concluded with unintelligible abruptness. Bruhle dies, purposely as it should seem that his pupil may renounce the romantic sublimity of his nature, and that his inauspicious union and prostituted character might be exempt from the censure of violated friendship. Numerous indications of profound and vigorous thought are scattered over even the most negligently compacted portions of the narrative. It is an unweeded garden where nightshade is interwoven with sweet jessamine, and the most delicate spices of the east peep over struggling stalks of rank and poisonous hemlock.

In the delineation of the more evanescent feelings and uncommon instances of strong and delicate passion we conceive the author to have exhibited new and unparalleled powers. He has noticed some peculiarities of female character with a delicacy and truth singularly exquisite. We think that the interesting subject of sexual relations requires for its successful development the application of a mind thus organized and endowed. Yet even here how great the deficiencies; this mind must be pure from the fashionable superstitions of gallantry, must be exempt from the sordid feelings which with blind idolatry worship the image and blaspheme the deity, reverence the type, and degrade the reality of which it is an emblem.

We do not hesitate to assert that the author of this volume is a man of ability. His great though indisciplinable energies and fervid rapidity of conception embody scenes and situations, and passions affording inexhaustible food for wonder and delight. The interest is deep and irresistible. A moral enchanter seems to have conjured up the shapes of all that is beautiful and strange to suspend the faculties in fascination and astonishment.

1814.

# A FABLE

## (TRANSLATION)

THERE was a youth who travelled through distant lands, seeking throughout the world a lady of whom he was enamoured. And who this lady was, and how this youth became enamoured of her, and how and why the great love he bore her forsook him, are things worthy to be known by every gentle heart.

At the dawn of the fifteenth spring of his life, a certain one calling himself Love awoke him, saying that one whom he had ofttimes beheld in his dreams abode awaiting him. This Love was accompanied by a great troop of female forms, all veiled in white, and crowned with laurel, ivy, and myrtle, garlanded and interwreathed with violets, roses, and lilies. They sang with such sweetness that perhaps the harmony of the spheres, to which the stars dance, is not so sweet. And their manners and words were so alluring, that the youth was enticed, and, arising from his couch, made himself ready to do all the pleasure of him who called himself Love ; at whose behest he followed him by lonely ways and deserts and caverns, until the whole troop arrived at a solitary wood, in a gloomy valley between two most lofty mountains, which valley was planted in the manner of a labyrinth, with pines, cypresses, cedars, and yews, whose shadows begot a mixture of delight and sadness. And in this wood the youth for a whole year followed the uncertain footsteps of this his companion and guide, as the moon follows the earth, save that there was no change in him, and nourished by the fruit of a certain tree which grew in the midst of the labyrinth—a food sweet and bitter at once, which being cold as ice to the lips, appeared fire in the veins. The veiled figures were continually around him, ministers and attendants obedient to his least gesture, and messengers between him and Love, when Love might leave him for a little on his other errands. But these figures, albeit executing his every other command with swiftness, never would unveil themselves to him, although he anxiously besought them; one only excepted, whose name was Life, and who had the fame of a potent enchantress. She was tall of person and beautiful, cheerful and easy in her manners, and richly adorned, and, as it seemed from her ready unveiling of herself, she wished well to this youth. But he soon perceived that she was more false than any Siren, for by her counsel Love abandoned him in this savage place, with only the company of these shrouded figures, who, by their obstinately remaining veiled, had always wrought him dread. And none can expound whether these figures were the spectres of his own dead thoughts, or the shadows of the living thoughts of Love. Then Life, haply ashamed of her deceit, concealed herself within the cavern of a certain sister of hers dwelling there; and Love, sighing, returned to his third heaven.

Scarcely had Love departed, when the masked forms, released from his government, unveiled themselves before the astonished youth. And for many days these figures danced around him whithersoever he went, alternately mocking and threatening him; and in the night while he reposed they defiled in long and slow procession before his couch, each more hideous and terrible than the other. Their horrible aspect and loathsome figure so overcame his heart with sadness that the fair heaven, covered with that shadow, clothed itself in clouds before his eyes ; and he wept so much that the herbs upon his path, fed with tears instead of dew, became pale and bowed like himself. Weary at length of this suffering, he came to the grot of the Sister of Life, herself also an enchantress, and found her sitting before a pale fire of perfumed wood, singing laments sweet in their melancholy, and weaving a white shroud, upon which his name was half wrought, with the obscure and imperfect beginning of a certain other name ; and he besought her to tell him her own, and she said, with a faint but sweet voice, 'Death.' And the youth said, 'O lovely Death, I pray thee to aid me against these hateful phantoms, companions of thy sister, which cease not to torment me.' And Death comforted him, and took his hand with a smile, and kissed his brow and cheek, so that every vein thrilled with joy and fear, and made him abide with her in a chamber of her cavern, whither, she said, it was against Destiny that the wicked companions of Life should ever come. The youth continually conversing with Death, and she, like-minded to a sister, caressing him and showing him every courtesy both in deed and word, he quickly became enamoured of her, and Life herself, far less any of her troop, seemed fair to him no longer : and his passion so overcame him, that upon his knees he prayed Death to love him as he loved her, and consent to do his pleasure. But Death said, 'Audacious that thou art, with whose desire has Death ever complied ? If thou lovedst me not, perchance I might love thee—beloved by thee, I hate thee and I fly thee.' Thus saying, she went forth from the cavern, and her dusky and ethereal form was soon lost amid the interwoven boughs of the forest.

From that moment the youth pursued the track of Death; and so mighty was the love that led him, that he had encircled the world and searched through all its regions, and many years were already spent, but sorrows rather than years had blanched his locks and withered the flower of his beauty, when he found himself upon the confines of the very forest from which his wretched wanderings had begun. He cast himself upon the grass and wept for many hours, so blinded by his tears that for much time he did not perceive that not all that bathed his face and his bosom were his own, but that a lady bowed behind him wept for pity of his weeping. And lifting up his eyes he saw her, and it seemed to him never to have beheld so glorious a vision, and he doubted much whether she were a human creature. And his love of Death was suddenly changed into hate and suspicion, for this new love was so potent that it overcame every other thought. This compassionate lady at first loved him for mere pity; but love grew up swiftly with compassion, and she loved for Love's own sake, no one beloved by her having need of pity any more. This was the lady in whose quest Love had led the youth through that gloomy labyrinth of error and suffering, haply for that he esteemed him unworthy of so much glory, and perceived him too weak to support such exceeding joy. After having somewhat dried their tears, the twain walked together in that same forest, until Death stood before them, and said, 'Whilst, O youth, thou didst love me, I hated thee, and now that thou hatest me, I love thee, and wish so well to thee and thy bride that in my kingdom, which thou mayest call Paradise, I have set apart a chosen spot, where ye may securely fulfil your happy loves.' And the lady, offended, and perchance somewhat jealous by reason of the past love of her spouse, turned her back upon Death, saying within herself, ‘ What would this lover of my husband who comes here to trouble us ? ’ and cried, ‘ Life ! Life ! ’ and Life came, with a gay visage, crowned with a rainbow, and clad in a various mantle of chameleon skin ; and Death went away weeping, and departing said with a sweet voice, ‘ Ye mistrust me, but I forgive ye, and await ye where ye needs must come, for I dwell with Love and Eternity, with whom the souls whose love is everlasting must hold communion ; then will ye perceive whether I have deserved your distrust. Meanwhile I commend ye to Life ; and, sister mine, I beseech thee, by the love of that Death with whom thou wert twin born, not to employ thy customary arts against these lovers, but content thee with the tribute thou hast already received of sighs and tears, which are thy wealth.’ The youth, mindful of how great evil she had wrought him in that wood, mistrusted Life ; but the lady, although she doubted, yet being jealous of Death, . . .

1820.

## THREE FRAGMENTS ON BEAUTY

Why is the reflection in that canal more beautiful than the objects it reflects ? The colours are more vivid, and yet blended with more harmony ; the openings from within into the soft and tender colours of the distant wood, and the intersection of the mountain lines, surpass and misrepresent truth.

The mountains sweep to the plain like waves that meet in a chasm—the olive woods are as green as a sea and are waving in the wind—the shadows of the clouds are spotting the bosoms of the hills—a heron comes sailing over me—a butterfly flits near—

at intervals the pines give forth their sweet and prolonged response to the wind—the myrtle bushes are in bud, and the soil beneath me is carpeted with odoriferous flowers.

It is sweet to feel the beauties of nature in every pulsation, in every nerve—but it is far sweeter to be able to express this feeling to one who loves you. To feel all that is divine in the green-robed earth and the starry sky is a penetrating yet vivid pleasure which, when it is over, presses like the memory of misfortune; but if you can express those feelings—if, secure of sympathy (for without sympathy it is worse than the taste of those apples whose core is as bitter ashes), if thus secure you can pour forth into another's most attentive ear the feelings by which you are entranced, there is an exultation of spirit in the utterance—a glory of happiness which far transcends all human transports, and seems to invest the soul as the saints are with light, with a halo untainted, holy, and undying.

1819.

## CRITICAL NOTICES OF THE SCULPTURE IN THE FLORENCE GALLERY

### ON THE NIOBE

Of all that remains to us of Greek antiquity, this figure is perhaps the most consummate personification of loveliness, with regard to its countenance, as that of the Venus of the Tribune is with regard to its entire form of woman. It is colossal; the size adds to its value; because it allows to the spectator the choice of a greater number of points of view, and affords him a more analytical one, in which to catch a greater number of the infinite modes of expression, of which any form approaching ideal beauty is necessarily composed. It is the figure of a mother in the act of sheltering, from some divine and inevitable peril, the last, we may imagine, of her surviving children.

The little creature, terrified, as we may conceive, at the strange destruction of all its kindred, has fled to its mother and is hiding its head in the folds of her robe, and casting back one arm, as in a passionate appeal for defence, where it never before could have been sought in vain. She is clothed in a thin tunic of delicate woof; and her hair is fastened on her head into a knot, probably by that mother whose care will never fasten it again. Niobe is enveloped in profuse drapery, a portion of which the left hand has gathered up, and is in the act of extending it over the child in the instinct of shielding her from what reason knows to be inevitable. The right (as the restorer has properly imagined,) is drawing up her daughter to her: and with that instinctive gesture, and by its gentle pressure, is encouraging the child to believe that it can give security. The countenance of Niobe is the consummation of feminine majesty and loveliness, beyond which the imagination scarcely doubts that it can conceive anything.

That masterpiece of the poetic harmony of marble expresses other feelings. There is embodied a sense of the inevitable and rapid destiny which is consummating around her, as if it were already over. It seems as if despair and beauty had combined, and produced nothing but the sublimity of grief. As the motions of the form expressed the instinctive sense of the possibility of protecting the child, and the accustomed and affectionate assurance that she would find an asylum within her arms, so reason and imagination speak in the countenance the certainty that no mortal defence is of avail. There is no terror in the countenance, only grief—deep, remediless grief. There is no anger :—of what avail is indignation against what is known to be omnipotent ? There is no selfish shrinking from personal pain—there is no panic at supernatural agency—there is no adverting to herself as herself : the calamity is mightier than to leave scope for such emotions.

Everything is swallowed up in sorrow : she is all tears ; her countenance, in assured expectation of the arrow piercing its last victim in her embrace, is fixed on her omnipotent enemy. The pathetic beauty of the expression of her tender, and inexhaustible, and unquenchable despair, is beyond the effect of sculpture. As soon as the arrow shall pierce her last tie upon earth, the fable that she was turned into stone, or dissolved into a fountain of tears, will be but a feeble emblem of the sadness of hopelessness, in which the few and evil years of her remaining life, we feel, must flow away.

It is difficult to speak of the beauty of the countenance, or to make intelligible in words, from what such astonishing loveliness results.

The head, resting somewhat backward upon the full and flowing contour of the neck, is as in the act of watching an event momently to arrive. The hair is delicately divided on the forehead, and a gentle beauty gleams from the broad and clear forehead, over which its strings are drawn. The face is of an oval fullness, and the features conceived with the daring of a sense of power. In this respect it resembles the careless majesty which Nature stamps upon the rare masterpieces of her creation, harmonizing them as it were from the harmony of the spirit within. Yet all this not only consists with, but is the cause of the subtlest delicacy of clear and tender beauty—the expression at once of innocence and sublimity of soul—of purity and strength—of all that which touches the most removed and divine of the chords that make music in our thoughts—of that which shakes with astonishment even the most superficial.

## THE MINERVA

The head is of the highest beauty. It has a close helmet, from which the hair, delicately parted on the forehead, half escapes. The attitude gives entire effect to the perfect form of the neck, and to that full and beautiful moulding of the lower part of the face and mouth, which is in living beings the seat of the expression of a simplicity and integrity of nature. Her face, upraised to heaven, is animated with a profound, sweet, and impassioned melancholy, with an earnest, and fervid, and disinterested pleading against some vast and inevitable wrong. It is the joy and poetry of sorrow making grief beautiful, and giving it that nameless feeling which, from the imperfection of language, we call pain, but which is not all pain, though a feeling which makes not only its possessor, but the spectator of it, prefer it to what is called pleasure, in which all is not pleasure. It is difficult to think that this head, though of the highest ideal beauty, is the head of Minerva, although the attributes and attitude of the lower part of the statue certainly suggest that idea. The Greeks rarely, in their representations of the characters of their gods,—unless we call the poetic enthusiasm of Apollo a mortal passion,—expressed the disturbance of human feeling; and here is deep and impassioned grief animating a divine countenance. It is, indeed, divine. Wisdom (which Minerva may be supposed to emblem) is pleading earnestly with Power,—and invested with the expression of that grief, because it must ever plead so vainly. The drapery of the statue, the gentle beauty of the feet, and the grace of the attitude, are what may be seen in many other statues belonging to that astonishing era which produced it; such a countenance is seen in few.

This statue happens to be placed on a pedestal, the subject of whose relief is in a spirit wholly the reverse. It was probably an altar to Bacchus—possibly a funeral urn. Under the festoons of fruits and flowers that grace the pedestal, the corners of which are ornamented with the skulls of goats, are sculptured some figures of Maenads under the inspiration of the god. Nothing can be conceived more wild and terrible than their gestures, touching, as they do, the verge of distortion, into which their fine limbs and lovely forms are thrown. There is nothing, however, that exceeds the possibility of nature, though it borders on its utmost line.

The tremendous spirit of superstition, aided by drunkenness, producing something beyond insanity, seems to have caught them in its whirlwinds, and to bear them over the earth, as the rapid volutions of a tempest have the ever-changing trunk of a water-spout, or as the torrent of a mountain river whirls the autumnal leaves resistlessly along in its full eddies. The hair, loose and floating, seems caught in the tempest of their own tumultuous motion; their heads are thrown back, leaning with a strange delirium upon their necks, and looking up to heaven whilst they totter and stumble even in the energy of their tempestuous dance.

One represents Agave with the head of Pentheus in one hand, and in the other a great knife; a second has a spear with its pine cone, which was the Thyrsus;

another dances with mad voluptuousness ; the fourth is beating a kind of tambourine.

This was indeed a monstrous superstition, even in Greece, where it was alone capable of combining ideal beauty and poetical and abstract enthusiasm with the wild errors from which it sprung. In Rome it had a more familiar, wicked, and dry appearance ; it was not suited to the severe and exact apprehensions of the Romans, and their strict morals were violated by it, and sustained a deep injury, little analogous to its effects upon the Greeks, who turned all things—superstition, prejudice, murder, madness—to beauty.

## ON THE VENUS CALLED ANADYOMINE

She has just issued from the bath, and yet is animated with the enjoyment of it.

She seems all soft and mild enjoyment, and the curved lines of her fine limbs flow into each other with a never-ending sinuosity of sweetness. Her face expresses a breathless, yet passive and innocent voluptuousness, free from affectation. Her lips, without the sublimity of lofty and impetuous passion, the grandeur of enthusiastic imagination of the Apollo of the Capitol, or the union of both, like the Apollo Belvidere, have the tenderness of arch, yet pure and affectionate desire, and the mode in which the ends of the mouth are drawn in, yet lifted or half-opened, with the smile that for ever circles round them, and the tremulous curve into which they are wrought by inextinguishable desire, and the tongue lying against the lower lip, as in the listlessness of passive joy, express love, still love.

Her eyes seem heavy and swimming with pleasure, and her small forehead fades on both sides into that sweet swelling and thin declension of the bone over the eye, in the mode which expresses simple and tender feelings.

The neck is full, and panting as with the aspiration of delight, and flows with gentle curves into her perfect form.

Her form is indeed perfect. She is half-sitting and half-rising from a shell, and the fullness of her limbs, and their complete roundness and perfection, do not diminish the vital energy with which they seem to be animated. The position of the arms, which are lovely beyond imagination, is natural, unaffected, and easy. This, perhaps, is the finest personification of Venus, the deity of superficial desire, in all antique statuary. Her pointed and pear-like person, ever virgin, and her attitude modesty itself.

## A BAS-RELIEF

(*Probably the sides of a Sarcophagus*)

The lady is lying on a couch, supported by a young woman, and looking extremely exhausted; her dishevelled hair is floating about her shoulder, and she is half-covered with drapery that falls on the couch.

Her tunic is exactly like a chemise, only the sleeves are longer, coming half-way down the upper part of the arm. An old wrinkled woman, with a cloak over her head, and an enormously sagacious look, has a most *professional* appearance, and is taking hold of her arm gently with one hand, and with the other is supporting it. I think she is feeling her pulse. At the side of the couch sits a woman as in grief, holding her head in her hands. At the bottom of the bed is another matron tearing her hair, and in the act of screaming out most violently, which she seems, however, by the rest of her gestures, to do with the utmost deliberation, as having come to the resolution, that it was a correct thing to do so. Behind her is a gossip of the most ludicrous ugliness, crying, I suppose, or praying, for her arms are crossed upon her neck. There is also a fifth setting up a wail. To the left of the couch a nurse is sitting on the ground dangling the child in her arms, and wholly occupied in so doing. The infant is swaddled. Behind her is a female who appears to be in the act of rushing in with dishevelled hair and violent gesture, and in one hand brandishing a whip or a thunder-bolt. This is probably some emblematic person, the messenger of death, or a fury, whose personification would be a key to the whole. What they are all wailing at, I know not; whether the lady is dying, or the father has directed the child to be exposed; but if the mother be not dead, such a tumult would kill a woman in the straw in these days.

The other compartment, in the second scene of the drama, tells the story of the presentation of the child to its father. An old man has it in his arms, and with professional and mysterious officiousness is holding it out to the father. The father, a middle-aged and very respectable-looking man, perhaps not long married, is looking with the admiration of a bachelor on his first child, and perhaps thinking, that he was once such a strange little creature himself. His hands are clasped, and he is gathering up between his arms the folds of his cloak, an emblem of his gathering up all his faculties to understand the tale the gossip is bringing.

An old man is standing beside him, probably his father, with some curiosity, and much tenderness in his looks. Around are collected a host of his relations, of whom the youngest, a handsome girl, seems the least concerned. It is altogether an admirable piece, quite in the spirit of the comedies of Terence.

## MICHAEL ANGELO'S BACCHUS

The countenance of this figure is a most revolting mistake of the spirit and meaning of Bacchus. It looks drunken, brutal, narrow-minded, and has an expression of dissoluteness the most revolting. The lower part of the figure is stiff, and the manner in which the shoulders are united to the breast, and the neck to the head, abundantly inharmonious. It is altogether without unity, as was the idea of the deity of Bacchus in the conception of a Catholic. On the other hand, considered only as a piece of workmanship, it has many merits. The arms are executed in a style of the most perfect and manly beauty. The body is conceived with great energy, and the manner in which the lines mingle into each other, of the highest boldness and truth. It wants unity as a work of art—as a representation of Bacchus it wants everything.

## A JUNO

A statue of great merit. The countenance expresses a stern and unquestioned severity of dominion, with a certain sadness. The lips are beautiful—susceptible of expressing scorn—but not without sweetness. With fine lips a person is never wholly bad, and they never belong to the expression of emotions wholly selfish—lips being the seat of imagination. The drapery is finely conceived, and the manner in which the act of throwing back one leg is expressed, in the diverging folds of the drapery of the left breast fading in bold yet graduated lines into a skirt, as it descends from the left shoulder, is admirably imagined.

AN APOLLO,

with serpents twining round a wreath of laurel on which the quiver is suspended. It probably was, when complete, magnificently beautiful. The restorer of the head and arms, following the indication of the muscles of the right side, has lifted the arm, as in triumph, at the success of an arrow, imagining to imitate the Lycian Apollo in that, so finely described by Apollonius Rhodius, when the dazzling radiance of his beautiful limbs shone over the dark Euxine. The action, energy, and godlike animation of these limbs speak a spirit which seems as if it could not be consumed.

1819.

# ESSAY ON THE LITERATURE, THE ARTS, AND THE MANNERS OF THE ATHENIANS

## A FRAGMENT

The period which intervened between the birth of Pericles and the death of Aristotle, is undoubtedly, whether considered in itself, or with reference to the effects which it has produced upon the subsequent destinies of civilized man, the most memorable in the history of the world. What was the combination of moral and political circumstances which produced so unparalleled a progress during that period in literature and the arts ;—why that progress, so rapid and so sustained, so soon received a check, and became retrograde,—are problems left to the wonder and conjecture of posterity. The wrecks and fragments of those subtle and profound minds, like the ruins of a fine statue, obscurely suggest to us the grandeur and perfection of the whole. Their very language—a type of the understandings of which it was the creation and the image—in variety, in simplicity, in flexibility, and in copiousness, excels every other language of the western world. Their sculptures are such as we, in our presumption, assume to be the models of ideal truth and beauty, and to which no artist of modern times can produce forms in any degree comparable. Their paintings, according to Pliny and Pausanias, were full of delicacy and harmony; and some even were powerfully pathetic, so as to awaken, like tender music or tragic poetry, the most overwhelming emotions. We are accustomed to conceive the painters of the sixteenth century, as those who have brought their art to the highest perfection, probably because none of the ancient paintings have been preserved. For all the inventive arts maintain, as it were, a sympathetic connexion between each other, being no more than various expressions of one internal power, modified by different circumstances, either of an individual, or of society; and the paintings of that period would probably bear the same relation as is confessedly borne by the sculptures to all succeeding ones. Of their music we know little; but the effects which it is said to have produced, whether they be attributed to the skill of the composer, or the sensibility of his audience, are far more powerful than any which we experience from the music of our own times; and if, indeed, the melody of their compositions were more tender and delicate, and inspiring, than the melodies of some modern European nations, their superiority in this art must have been something wonderful, and wholly beyond conception.

Their poetry seems to maintain a very high, though not so disproportionate a rank, in the comparison.

Perhaps Shakespeare, from the variety and comprehension of his genius, is to be considered, on the whole, as the greatest individual mind, of which we have specimens remaining. Perhaps Dante created imaginations of greater loveliness and energy than any that are to be found in the ancient literature of Greece. Perhaps nothing has been discovered in the fragments of the Greek lyric poets equivalent to the sublime and chivalric sensibility of Petrarch.—But, as a poet, Homer must be acknowledged to excel Shakespeare in the truth, the harmony, the sustained grandeur, the satisfying completeness of his images, their exact fitness to the illustration, and to that to which they belong. Nor could Dante, deficient in conduct, plan, nature, variety, and temperance, have been brought into comparison with these men, but for those fortunate isles, laden with golden fruit, which alone could tempt any one to embark in the misty ocean of his dark and extravagant fiction.

But, omitting the comparison of individual minds, which can afford no general inference, how superior was the spirit and system of their poetry to that of any other period! So that, had any other genius equal in other respects to the greatest that ever enlightened the world, arisen in that age, he would have been superior to all, from this circumstance alone—that his conceptions would have assumed a more harmonious and perfect form. For it is worthy of observation, that whatever the poets of that age produced is as harmonious and perfect as possible. If a drama, for instance, were the composition of a person of inferior talent, it was still homogeneous and free from inequalities ; it was a whole, consistent with itself. The compositions of great minds bore throughout the sustained stamp of their greatness. In the poetry of succeeding ages the expectations are often exalted on Icarian wings, and fall, too much disappointed to give a memory and a name to the oblivious pool in which they fell.

In physical knowledge Aristotle and Theophrastus had already—no doubt assisted by the labours of those of their predecessors whom they criticize—made advances worthy of the maturity of science. The astonishing invention of geometry, that series of discoveries which have enabled man to command the elements and foresee future events, before the subjects of his ignorant wonder, and which have opened as it were the doors of the mysteries of nature, had already been brought to great perfection. Metaphysics, the science of man's intimate nature, and logic, or the grammar and elementary principles of that science, received from the latter philosophers of the Periclean age a firm basis. All our more exact philosophy is built upon the labours of these great men, and many of the words which we employ in metaphysical distinctions were invented by them to give accuracy and system to their reasonings. The science of morals, or the voluntary conduct of men in relation to themselves or others, dates from this epoch. How inexpressibly bolder and more pure were the doctrines of those great men, in comparison with the timid maxims which prevail in the writings of the most esteemed modern moralists! They were such as Phocion, and Epaminondas, and Timoleon, who formed themselves on their influence, were to the wretched heroes of our own age.

Their political and religious institutions are more difficult to bring into comparison with those of other times. A summary idea may be formed of the worth of any political and religious system, by observing the comparative degree of happiness and of intellect produced under its influence. And whilst many institutions and opinions, which in ancient Greece were obstacles to the improvement of the human race, have been abolished among modern nations, how many pernicious superstitions and new contrivances of misrule, and unheard-of complications of public mischief, have not been invented among them by the ever-watchful spirit of avarice and tyranny!

The modern nations of the civilized world owe the progress which they have made—as well in those physical sciences in which they have already excelled their masters, as in the moral and intellectual inquiries, in which, with all the advantage of the experience of the latter, it can scarcely be said that they have yet equalled them,—to what is called the revival of learning; that is, the study of the writers of the age which preceded and immediately followed the government of Pericles, or of subsequent writers, who were, so to speak, the rivers flowing from those immortal fountains. And though there seems to be a principle in the modern world, which, should circumstances analogous to those which modelled the intellectual resources of the age to which we refer, into so harmonious a proportion, again arise, would arrest and perpetuate them, and consign their results to a more equal, extensive, and lasting improvement of the condition of man—though justice and the true meaning of human society are, if not more accurately, more generally understood; though perhaps men know more, and therefore are more, as a mass, yet this principle has never been called into action, and requires indeed a universal and an almost appalling change in the system of existing things. The study of modern history is the study of kings, financiers, statesmen, and priests. The history of ancient Greece is the study of legislators, philosophers, and poets; it is the history of men, compared with the history of titles. What the Greeks were, was a reality, not a promise. And what we are and hope to be, is derived, as it were, from the influence and inspiration of these glorious generations.

Whatever tends to afford a further illustration of the manners and opinions of those to whom we owe so much, and who were perhaps, on the whole, the most perfect specimens of humanity of whom we have authentic record, were infinitely valuable. Let us see their errors, their weaknesses, their daily actions, their familiar conversation, and catch the tone of their society. When we discover how far the most admirable community ever framed was removed from that perfection to which human society is impelled by some active power within each bosom to aspire, how great ought to be our hopes, how resolute our struggles! For the Greeks of the Periclean age were widely different from us. It is to be lamented that no modern writer has hitherto dared to show them precisely as they were. Barthélemi cannot be denied the praise of industry and system; but he never forgets that he is a Christian and a Frenchman. Wieland, in his delightful novels, makes indeed a very tolerable Pagan, but cherishes too many political prejudices, and refrains from diminishing the interest of his romances by painting sentiments in which no European of modern times can possibly sympathize. There is no book which shows the Greeks precisely as they were; they seem all written for children, with the caution that no practice or sentiment, highly inconsistent with our present manners, should be mentioned, lest those manners should receive outrage and violation. But there are many to whom the Greek language is inaccessible, who ought not to be excluded by this prudery from possessing an exact and comprehensive conception of the history of man;

for there is no knowledge concerning what man has been and may be, from partaking of which a person can depart, without becoming in some degree more philosophical, tolerant, and just.

One of the chief distinctions between the manners of ancient Greece and modern Europe, consisted in the regulations and the sentiments respecting sexual intercourse. Whether this difference arises from some imperfect influence of the doctrines of Jesus, who alleges the absolute and unconditional equality of all human beings, or from the institutions of chivalry, or from a certain fundamental difference of physical nature existing in the Celts, or from a combination of all or any of these causes acting on each other, is a question worthy of voluminous investigation. The fact is, that the modern Europeans have in this circumstance, and in the abolition of slavery, made an improvement the most decisive in the regulation of human society; and all the virtue and the wisdom of the Periclean age arose under other institutions, in spite of the diminution which personal slavery and the inferiority of women, recognized by law and opinion, must have produced in the delicacy, the strength, the comprehensiveness, and the accuracy of their conceptions, in moral, political, and metaphysical science, and perhaps in every other art and science.

The women, thus degraded, became such as it was expected they would become. They possessed, except with extraordinary exceptions, the habits and the qualities of slaves. They were probably not extremely beautiful; at least there was no such disproportion in the attractions of the external form between the female and male sex among the Greeks, as exists among the modern Europeans. They were certainly devoid of that moral and intellectual loveliness with which the acquisition of knowledge and the cultivation of sentiment animates, as with another life of overpowering grace, the lineaments and the gestures of every form which they inhabit. Their eyes could not have been deep and intricate from the workings of the mind, and could have entangled no heart in soul-enwoven labyrinths.

Let it not be imagined that because the Greeks were deprived of its legitimate object, they were incapable of sentimental love; and that this passion is the mere child of chivalry and the literature of modern times. This object or its archetype for ever exists in the mind, which selects among those who resemble it that which most resembles it; and instinctively fills up the interstices of the imperfect image, in the same manner as the imagination moulds and completes the shapes in clouds, or in the fire, into the resemblances of whatever form, animal, building, &c., happens to be present to it. Man is in his wildest state a social being: a certain degree of civilization and refinement ever produces the want of sympathies still more intimate and complete; and the gratification of the senses is no longer all that is sought in sexual connexion. It soon becomes a very small part of that profound and complicated sentiment, which we call love, which is rather the universal thirst for a communion not only of the senses, but of our whole nature, intellectual, imaginative and sensitive, and which, when individualized, becomes an imperious necessity, only to be satisfied by the complete or partial, actual or supposed fulfilment of its claims. This want grows more powerful in proportion to the development which our nature receives from civilization, for man never ceases to be a social being. The sexual impulse, which is only one, and often a small part of those claims, serves, from its obvious and external nature, as a kind of type or expression of the rest, a common basis, an acknowledged and visible link. Still it is a claim which even derives a strength not its own from the accessory circumstances which surround it, and one which our nature thirsts to satisfy. To estimate this, observe the degree of intensity and durability of the love of the male towards the female in animals and savages; and acknowledge all the duration and intensity observable in the love of civilized beings beyond that of savages to be produced from other causes. In the susceptibility of the external senses there is probably no important difference.

Among the ancient Greeks the male sex, one half of the human race, received the highest cultivation and refinement: whilst the other, so far as intellect is concerned, were educated as slaves, and were raised but few degrees in all that related to moral or intellectual excellence above the condition of savages. The gradations in the society of man present us with slow improvement in this respect. The Roman women held a higher consideration in society, and were esteemed almost as the equal partners with their husbands in the regulation of domestic economy and the education of their children. The practices and customs of modern Europe are essentially different from and incomparably less pernicious than either, however remote from what an enlightened mind cannot fail to desire as the future destiny of human beings.

1818

# ON THE SYMPOSIUM, OR PREFACE TO THE BANQUET OF PLATO

### A FRAGMENT

THE dialogue entitled *The Banquet* was selected by the translator as the most beautiful and perfect among all the works of Plato.[1] He despairs of having communicated to the English language any portion of the surpassing graces of the composition, or having done more than present an imperfect shadow of the language and the sentiment of this astonishing production.

Plato is eminently the greatest among the Greek philosophers, and from, or, rather, perhaps through him, his master Socrates, have proceeded those emanations of moral and metaphysical knowledge, on which a long series and an incalculable variety of popular superstitions have sheltered their absurdities from the slow contempt of mankind. Plato exhibits the rare union of close and subtle logic with the Pythian enthusiasm of poetry, melted by the splendour and harmony of his periods into one irresistible stream of musical impressions, which hurry the persuasions onward, as in a breathless career. His language is that of an immortal spirit, rather than a man. Lord Bacon is, perhaps, the only writer, who, in these particulars, can be compared with him: his imitator, Cicero, sinks in the comparison into an ape mocking

[1] The *Republic*, though replete with considerable errors of speculation, is, indeed, the greatest repository of important truths of all the works of Plato. This, perhaps, is because it is the longest. He first, and perhaps last, maintained that a state ought to be governed, not by the wealthiest, or the most ambitious, or the most cunning, but by the wisest; the method of selecting such rulers, and the laws by which such a selection is made, must correspond with and arise out of the moral freedom and refinement of the people.

the gestures of a man. His views into the nature of mind and existence are often obscure, only because they are profound; and though his theories respecting the government of the world, and the elementary laws of moral action, are not always correct, yet there is scarcely any of his treatises which do not, however stained by puerile sophisms, contain the most remarkable intuitions into all that can be the subject of the human mind. His excellence consists especially in intuition, and it is this faculty which raises him far above Aristotle, whose genius, though vivid and various, is obscure in comparison with that of Plato.

The dialogue entitled the *Banquet*, is called Ερωτικος, or a Discussion upon Love, and is supposed to have taken place at the house of Agathon, at one of a series of festivals given by that poet, on the occasion of his gaining the prize of tragedy at the Dionysiaca. The account of the debate on this occasion is supposed to have been given by Apollodorus, a pupil of Socrates, many years after it had taken place, to a companion who was curious to hear it. This Apollodorus appears, both from the style in which he is represented in this piece, as well as from a passage in the *Phaedon*, to have been a person of an impassioned and enthusiastic disposition; to borrow an image from the Italian painters, he seems to have been the St. John of the Socratic group. The drama (for so the lively distinction of character and the various and well-wrought circumstances of the story almost entitle it to be called) begins by Socrates persuading Aristodemus to sup at Agathon's, uninvited. The whole of this introduction affords the most lively conception of refined Athenian manners.

1818

[UNFINISHED]

# ON LOVE

What is love ? Ask him who lives, what is life ? ask him who adores, what is God ?

I know not the internal constitution of other men, nor even thine, whom I now address. I see that in some external attributes they resemble me, but when, misled by that appearance, I have thought to appeal to something in common, and unburthen my inmost soul to them, I have found my language misunderstood, like one in a distant and savage land. The more opportunities they have afforded me for experience, the wider has appeared the interval between us, and to a greater distance have the points of sympathy been withdrawn. With a spirit ill fitted to sustain such proof, trembling and feeble through its tenderness, I have everywhere sought sympathy and have found only repulse and disappointment.

*Thou* demandest what is love ? It is that powerful attraction towards all that we conceive, or fear, or hope beyond ourselves, when we find within our own thoughts the chasm of an insufficient void, and seek to awaken in all things that are, a community with what we experience within ourselves. If we reason, we would be understood ; if we imagine, we would that the airy children of our brain were born anew within another's ; if we feel, we would that another's nerves should vibrate to our own, that the beams of their eyes should kindle at once and mix and melt into our own, that lips of motionless ice should not reply to lips quivering and burning with the heart's best blood. This is Love. This is the bond and the sanction which connects not only man with man, but with everything which exists. We are born into the world, and there is something within us which, from the instant that we live, more and more thirsts after its likeness. It is probably in correspondence with this law that the infant drains milk from the bosom of its mother ; this propensity develops itself with the development of our nature. We dimly see within our intellectual nature a miniature as it were of our entire self, yet deprived of all that we condemn or despise, the ideal prototype of everything excellent or lovely that we are capable of conceiving as belonging to the nature of man. Not only the portrait of our external being, but an assemblage of the minutest particles of which our nature is composed ;[1] a mirror whose surface reflects only the forms of purity and brightness ; a soul within our soul that describes a circle around its proper paradise, which pain, and sorrow, and evil dare not overleap. To this we eagerly refer all sensations, thirsting that they should resemble or correspond with it. The discovery of its antitype ; the meeting with an understanding capable of clearly estimating our own ; an imagination which should enter into and seize upon the subtle and delicate peculiarities which we have delighted to cherish and unfold in secret ; with a frame whose nerves, like the chords of two exquisite lyres, strung to the accompaniment of one delightful voice, vibrate with the vibrations of our own ; and of a combination of all these in such proportion as the type within demands ; this is the invisible and unattainable point to which Love tends ; and to attain which, it urges forth the powers of man to arrest the faintest shadow of that, without the possession of which there is no rest nor respite to the heart over which it rules. Hence in solitude, or in that deserted state when we are surrounded by human beings, and yet they sympathize

[1] These words are ineffectual and metaphorical. Most words are so—No help !

not with us, we love the flowers, the grass, and the waters, and the sky. In the motion of the very leaves of spring, in the blue air, there is then found a secret correspondence with our heart. There is eloquence in the tongueless wind, and a melody in the flowing brooks and the rustling of the reeds beside them, which by their inconceivable relation to something within the soul, awaken the spirits to a dance of breathless rapture, and bring tears of mysterious tenderness to the eyes, like the enthusiasm of patriotic success, or the voice of one beloved singing to you alone. Sterne says that, if he were in a desert, he would love some cypress. So soon as this want or power is dead, man becomes the living sepulchre of himself, and what yet survives is the mere husk of what once he was.

1815

# ON THE PUNISHMENT OF DEATH

### A FRAGMENT

THE first law which it becomes a Reformer to propose and support, at the approach of a period of great political change, is the abolition of the punishment of death.

It is sufficiently clear that revenge, retaliation, atonement, expiation, are rules and motives, so far from deserving a place in any enlightened system of political life, that they are the chief sources of a prodigious class of miseries in the domestic circles of society. It is clear that however the spirit of legislation may appear to frame institutions upon more philosophical maxims, it has hitherto, in those cases which are termed criminal, done little more than palliate the spirit, by gratifying a portion of it; and afforded a compromise between that which is best ;—the inflicting of no evil upon a sensitive being, without a decisively beneficial result in which he should at least participate ;—and that which is worst ; that he should be put to torture for the amusement of those whom he may have injured, or may seem to have injured.

Omitting these remoter considerations, let us inquire what *Death* is ; that punishment which is applied as a measure of transgressions of indefinite shades of distinction, so soon as they shall have passed that degree and colour of enormity, with which it is supposed no inferior infliction is commensurate.

And first, whether death is good or evil, a punishment or a reward, or whether it be wholly indifferent, no man can take upon himself to assert. That that within us which thinks and feels, continues to think and feel after the dissolution of the body, has been the almost universal opinion of mankind, and the accurate philosophy of what I may be permitted to term the modern Academy, by showing the prodigious depth and extent of our ignorance respecting the causes and nature of sensation, renders probable the affirmative of a proposition, the negative of which it is so difficult to conceive, and the popular arguments against which, derived from what is called the atomic system, are proved to be applicable only to the relation which one object bears to another, as apprehended by the mind, and not to existence itself, or the nature of that essence which is the medium and receptacle of objects.

The popular system of religion suggests the idea that the mind, after death, will be painfully or pleasurably affected according to its determinations during life. However ridiculous and pernicious we must admit the vulgar accessories of this creed to be, there is a certain analogy, not wholly absurd, between the consequences resulting to an individual during life from the virtuous or vicious, prudent or imprudent, conduct of his external actions, to those consequences which are conjectured to ensue from the discipline and order of his internal thoughts, as affecting his condition in a future state. They omit, indeed, to calculate upon the accidents of disease, and temperament, and organization, and circumstance, together with the multitude of independent agencies which affect the opinions, the conduct, and the happiness of individuals, and produce determinations of the will, and modify the judgement, so as to produce effects the most opposite in natures considerably similar. These are those operations in the order of the whole of nature, tending, we are prone to believe, to some definite mighty end, to which the agencies of our peculiar nature are subordinate ; nor is there any reason to suppose, that in a future state they should become suddenly exempt from that subordination. The philosopher is unable to determine whether our existence in a previous state has affected our present condition, and abstains from deciding whether our present condition will affect us in that which may be future. That, if we continue to exist, the manner of our existence will be such as no inferences nor conjectures, afforded by a consideration of our earthly experience, can elucidate, is sufficiently obvious. The opinion that the vital principle within us, in whatever mode it may continue to exist, must lose that consciousness of definite and individual being which now characterizes it, and become a unit in the vast sum of action and of thought which disposes and animates the universe, and is called God, seems to belong to that class of opinion which has been designated as indifferent.

To compel a person to know all that can be known by the dead, concerning that which the living fear, hope, or forget; to plunge him into the pleasure or pain which there awaits him; to punish or reward him in a manner and in a degree incalculable and incomprehensible by us; to disrobe him at once from all that intertexture of good and evil with which Nature seems to have clothed every form of individual existence, is to inflict on him the doom of death.

A certain degree of pain and terror usually accompany the infliction of death. This degree is infinitely varied by the infinite variety in the temperament and opinions of the sufferers. As a measure of punishment, strictly so considered, and as an exhibition, which, by its known effects on the sensibility of the sufferer, is intended to intimidate the spectators from incurring a similar liability, it is singularly inadequate.

Firstly,—Persons of energetic character, in whom, as in men who suffer for political crimes, there is a large mixture of enterprise, and fortitude, and disinterestedness, and the elements, though misguided and disarranged, by which the strength and happiness of a nation might have been cemented, die in such a manner, as to make death appear not evil, but good. The death of what is called a traitor, that is, a person who, from whatever motive, would abolish the government of the day, is as often a triumphant exhibition of suffering virtue, as the warning of a culprit. The multitude, instead of departing with a panic-stricken approbation of the laws which exhibited such a spectacle, are inspired with pity, admiration and sympathy; and the most generous among them feel an emulation to be the authors of such flattering emotions, as they experience stirring in their bosoms. Impressed by what they see and feel, they make no distinction between the motives which incited the criminals to the actions for which they suffer, or the heroic courage with which they turned into good that which their judges awarded to them as evil, or the purpose itself of those actions, though that purpose may happen to be eminently pernicious. The laws in this case lose that sympathy, which it ought to be their chief object to secure, and in a participation of which consists their chief strength in maintaining those sanctions by which the parts of the social union are bound together, so as to produce, as nearly as possible, the ends for which it is instituted.

Secondly,—Persons of energetic character, in communities not modelled with philosophical skill to turn all the energies which they contain to the purposes of common good, are prone also to fall into the temptation of undertaking, and are peculiarly fitted for despising the perils attendant upon consummating, the most enormous crimes. Murder, rapes, extensive schemes of plunder, are the actions of persons belonging to this class; and death is the penalty of conviction. But the coarseness of organization, peculiar to men capable of committing acts wholly selfish, is usually found to be associated with a proportionate insensibility to fear or pain. Their sufferings communicate to those of the spectators, who may be liable to the commission of similar crimes, a sense of the lightness of that event, when closely examined, which, at a distance, as uneducated persons are accustomed to do, probably they regarded with horror. But a great majority of the spectators are so bound up in the interests and the habits of social union that no temptation would be sufficiently strong to induce them to a commission of the enormities to which this penalty is assigned. The more powerful, and the richer among them,—and a numerous class of little tradesmen are richer and more powerful than those who are employed by them, and the employer, in general, bears this relation to the employed,—regard their own wrongs as, in some degree, avenged, and their own rights secured by this punishment, inflicted as the penalty of whatever crime. In cases of murder or mutilation, this feeling is almost universal. In those, therefore, whom this exhibition does not awaken to the sympathy which extenuates crime and discredits the law which restrains it, it produces feelings more directly at war with the genuine purposes of political society. It excites those emotions which it is the chief object of civilization to extinguish for ever, and in the extinction of which alone there can be any hope of better institutions than those under which men now misgovern one another. Men feel that their revenge is gratified, and that their security is established by the extinction and the sufferings of beings, in most respects resembling themselves; and their daily occupations constraining them to a precise form in all their thoughts, they come to connect inseparably the idea of their own advantage with that of the death and torture of others. It is manifest that the object of sane polity is directly the reverse; and that laws founded upon reason, should accustom the gross vulgar to associate their ideas of security and of interest with the reformation, and the strict restraint, for that purpose alone, of those who might invade it.

The passion of revenge is originally nothing more than an habitual perception of the ideas of the sufferings of the person who inflicts an injury, as connected, as they are in a savage state, or in such portions of society as are yet undisciplined to civilization, with security that that injury will not be repeated in future. This feeling, engrafted upon superstition and confirmed by habit, at last loses sight of the only object for which it may be supposed to have been implanted, and becomes a passion and a duty to be pursued and fulfilled, even to the destruction of those ends to which it originally tended. The other passions, both good and evil, Avarice, Remorse, Love, Patriotism, present a similar appearance; and to this principle of the mind over-shooting the mark at which it aims, we owe all that is eminently base or excellent in human nature ; in providing for the nutriment or the extinction of which, consists the true art of the legislator.[1]

Nothing is more clear than that the infliction of punishment in general, in a degree which the reformation and the restraint of those who transgress the laws does not render indispensable, and none more than death, confirms all the inhuman and unsocial impulses of men. It is almost a proverbial remark, that those nations in which the penal code has been particularly mild, have been distinguished from all others by the rarity of crime. But the example is to be admitted to be equivocal. A more decisive argument is afforded by a consideration of the universal connexion of ferocity of manners, and a contempt of social ties, with the contempt of human life. Governments which derive their institutions from the exis-

[1] The savage and the illiterate are but faintly aware of the distinction between the future and the past ; they make actions belonging to periods so distinct, the subjects of similar feelings ; they live only in the present, or in the past, as it is present. It is in this that the philosopher excels one of the many ; it is this which distinguishes the doctrine of philosophic necessity from fatalism ; and that determination of the will, by which it is the active source of future events, from that liberty or indifference, to which the abstract liability of irremediable actions is attached, according to the notions of the vulgar.

This is the source of the erroneous excesses of Remorse and Revenge ; the one extending itself over the future, and the other over the past ; provinces in which their suggestions can only be the sources of evil. The purpose of a resolution to act more wisely and virtuously in future, and the sense of a necessity of caution in repressing an enemy, are the sources from which the enormous superstitions implied in the words cited have arisen.

tence of circumstances of barbarism and violence, with some rare exceptions perhaps, are bloody in proportion as they are despotic, and form the manners of their subjects to a sympathy with their own spirit.

The spectators who feel no abhorrence at a public execution, but rather a self-applauding superiority, and a sense of gratified indignation, are surely excited to the most inauspicious emotions. The first reflection of such a one is the sense of his own internal and actual worth, as preferable to that of the victim, whom circumstances have led to destruction. The meanest wretch is impressed with a sense of his own comparative merit. He is one of those on whom the tower of Siloam fell not—he is such a one as Jesus Christ found not in all Samaria, who, in his own soul, throws the first stone at the woman taken in adultery. The popular religion of the country takes its designation from that illustrious person whose beautiful sentiment I have quoted. Any one who has stript from the doctrines of this person the veil of familiarity, will perceive how adverse their spirit is to feelings of this nature.

1815

# ON LIFE

Life and the world, or whatever we call that which we are and feel, is an astonishing thing. The mist of familiarity obscures from us the wonder of our being. We are struck with admiration at some of its transient modifications, but it is itself the great miracle. What are changes of empires, the wreck of dynasties, with the opinions which supported them ; what is the birth and the extinction of religious and of political systems to life ? What are the revolutions of the globe which we inhabit, and the operations of the elements of which it is composed, compared with life ? What is the universe of stars, and suns, of which this inhabited earth is one, and their motions, and their destiny, compared with life? Life, the great miracle, we admire not, because it is so miraculous. It is well that we are thus shielded by the familiarity of what is at once so certain and so unfathomable, from an astonishment which would otherwise absorb and overawe the functions of that which is its object.

If any artist, I do not say had executed, but had merely conceived in his mind the system of the sun, and the stars, and planets, they not existing, and had painted to us in words, or upon canvas, the spectacle now afforded by the nightly cope of heaven, and illustrated it by the wisdom of astronomy, great would be our admiration. Or had he imagined the scenery of this earth, the mountains, the seas, and the rivers; the grass, and the flowers, and the variety of the forms and masses of the leaves of the woods, and the colours which attend the setting and the rising sun, and the hues of the atmosphere, turbid or serene, these things not before existing, truly we should have been astonished, and it would not have been a vain boast to have said of such a man, 'Non merita nome di creatore, se non Iddio ed il Poeta.' But now these things are looked on with little wonder, and to be conscious of them with intense delight is esteemed to be the distinguishing mark of a refined and extraordinary person. The multitude of men care not for them. It is thus with Life—that which includes all.

What is life? Thoughts and feelings arise, with or without our will, and we employ words to express them. We are born, and our birth is unremembered, and our infancy remembered but in fragments; we live on, and in living we lose the apprehension of life. How vain is it to think that words can penetrate the mystery of our being! Rightly used they may make evident our ignorance to ourselves, and this is much. For what are we ? Whence do we come ? and whither do we go ? Is birth the commencement, is death the conclusion of our being ? What is birth and death ?

The most refined abstractions of logic conduct to a view of life, which, though startling to the apprehension, is, in fact, that which the habitual sense of its repeated combinations has extinguished in us. It strips, as it were, the painted curtain from this scene of things. I confess that I am one of those who am unable to refuse my assent to the conclusions of those philosophers who assert that nothing exists but as it is perceived.

It is a decision against which all our persuasions struggle, and we must be long convicted before we can be convinced that the solid universe of external things is 'such stuff as dreams are made of'. The shocking absurdities of the popular philosophy of mind and matter, its fatal consequences in morals, and their violent dogmatism concerning the source of all things, had early conducted me to materialism. This materialism is a seducing system to young and superficial minds. It allows its disciples to talk, and dispenses them from thinking. But I was discontented with such a view of things as it afforded ; man is a being of high aspirations, 'looking both before and after,' whose 'thoughts wander through eternity', disclaiming alliance with transience and decay ; incapable of imagining to himself annihilation ; existing but in the future and the past ; being, not what he is, but what he has been and shall be. Whatever may be his true and final destination, there is a spirit within him at enmity with nothingness and dissolution. This is the character of all life and being. Each is at once the centre and the circumference ; the point to which all things are referred, and the line in which all things are contained. Such contemplations as these, materialism and the popular philosophy of mind and matter alike forbid ; they are only consistent with the intellectual system.

It is absurd to enter into a long recapitulation of arguments sufficiently familiar to those inquiring minds, whom alone a writer on abstruse subjects can be conceived to address. Perhaps the most clear and vigorous statement of the intellectual system is to be found in Sir William Drummond's Academical Questions. After such an exposition, it would be idle to translate into other words what could only lose its energy and fitness by the change. Examined point by point, and word by word, the most discriminating intellects have been able to discern no train of thoughts in the process of reasoning, which does not conduct inevitably to the conclusion which has been stated.

What follows from the admission ? It establishes no new truth, it gives us no additional insight into our hidden nature, neither its action nor itself. Philosophy, impatient as it may be to build, has much work yet remaining, as pioneer for the overgrowth of ages. It makes one step towards this object ; it destroys error, and the roots of error. It leaves, what it is too often the duty of the reformer in political and ethical questions to leave, a vacancy. It reduces the mind to that freedom in which it would have acted, but for the misuse of words and signs, the instruments of its own creation. By signs, I would be understood in a wide sense, including what is properly meant by that term, and what I peculiarly mean. In this latter sense, almost all familiar objects are signs, standing, not for themselves, but for others, in their capacity of suggesting one thought which shall lead to a train of thoughts. Our whole life is thus an education of error.

Let us recollect our sensations as children. What a distinct and intense apprehension had we of the world and of ourselves! Many of the circumstances of social life were then important to us which are now no longer so. But that is not the point of comparison on which I mean to insist. We less habitually distinguished all that we saw and felt, from ourselves. They seemed as it were to constitute one mass. There are some persons who, in this respect, are always children. Those who are subject to the state called reverie, feel as if their nature were dissolved into the surrounding universe, or as if the surrounding universe were absorbed into their being. They are conscious of no distinction. And these are states which precede, or accompany, or follow an unusually intense and vivid apprehension of life. As men grow up this power commonly decays, and they become mechanical and habitual agents. Thus feelings and then reasonings are the combined result of a multitude of entangled thoughts, and of a series of what are called impressions, planted by reiteration.

The view of life presented by the most refined deductions of the intellectual philosophy, is that of unity. Nothing exists but as it is perceived. The difference is merely nominal between those two classes of thought, which are vulgarly distinguished by the names of ideas and of external objects. Pursuing the same thread of reasoning, the existence of distinct individual minds, similar to that which is employed in now questioning its own nature, is likewise found to be a delusion. The words *I*, *you*, *they*, are not signs of any actual difference subsisting between the assemblage of thoughts thus indicated, but are merely marks employed to denote the different modifications of the one mind.

Let it not be supposed that this doctrine conducts to the monstrous presumption that I, the person who now write and think, am that one mind. I am but a portion of it. The words *I*, and *you*, and *they*, are grammatical devices invented simply for arrangement, and totally devoid of the intense and exclusive sense usually attached to them. It is difficult to find terms adequate to express so subtle a conception as that to which the Intellectual Philosophy has conducted us. We are on that verge where words abandon us, and what wonder if we grow dizzy to look down the dark abyss of how little we know.

The relations of *things* remain unchanged, by whatever system. By the word *things* is to be understood any object of thought, that is any thought upon which any other thought is employed, with an apprehension of distinction. The relations of these remain unchanged; and such is the material of our knowledge.

What is the cause of life ? that is, how was it produced, or what agencies distinct from life have acted or act upon life ? All recorded generations of mankind have weariedly busied themselves in inventing answers to this question ; and the result has been,—Religion. Yet, that the basis of all things cannot be, as the popular philosophy alleges, mind, is sufficiently evident. Mind, as far as we have any experience of its properties, and beyond that experience how vain is argument ! cannot create, it can only perceive. It is said also to be the cause. But cause is only a word expressing a certain state of the human mind with regard to the manner in which two thoughts are apprehended to be related to each other. If any one desires to know how unsatisfactorily the popular philosophy employs itself upon this great question, they need only impartially reflect upon the manner in which thoughts develop themselves in their minds. It is infinitely improbable that the cause of mind, that is, of existence, is similar to mind.

1815

# ON A FUTURE STATE

It has been the persuasion of an immense majority of human beings in all ages and nations that we continue to live after death,—that apparent termination of all the functions of sensitive and intellectual existence. Nor has mankind been contented with supposing that species of existence which some philosophers have asserted; namely, the resolution of the component parts of the mechanism of a living being into its elements, and the impossibility of the minutest particle of these sustaining the smallest diminution. They have clung to the idea that sensibility and thought, which they have distinguished from the objects of it, under the several names of spirit and matter, is, in its own nature, less susceptible of division and decay, and that, when the body is resolved into its elements, the principle which animated it will remain perpetual and unchanged. Some philosophers—and those to whom we are indebted for the most stupendous discoveries in physical science, suppose, on the other hand, that intelligence is the mere result of certain combinations among the particles of its objects; and those among them who believe that we live after death, recur to the interposition of a supernatural power, which shall overcome the tendency inherent in all material combinations, to dissipate and be absorbed into other forms.

Let us trace the reasonings which in one and the other have conducted to these two opinions, and endeavour to discover what we ought to think on a question of such momentous interest. Let us analyse the ideas and feelings which constitute the contending beliefs, and watchfully establish a discrimination between words and thoughts. Let us bring the question to the test of experience and fact; and ask ourselves, considering our nature in its entire extent, what light we derive from a sustained and comprehensive view of its component parts, which may enable us to assert, with certainty, that we do or do not live after death.

The examination of this subject requires that it should be stript of all those accessory topics which adhere to it in the common opinion of men. The existence of a God, and a future state of rewards and punishments, are totally foreign to the subject. If it be proved that the world is ruled by a Divine Power, no inference necessarily can be drawn from that circumstance in favour of a future state. It has been asserted, indeed, that as goodness and justice are to be numbered among the attributes of the Deity, He will undoubtedly compensate the virtuous who suffer during life, and that He will make every sensitive being, who does not deserve punishment, happy for ever. But this view of the subject, which it would be tedious as well as superfluous to develop and expose, satisfies no person, and cuts the knot which we now seek to untie. Moreover, should it be proved, on the other hand, that the mysterious principle which regulates the proceedings of the universe, is neither intelligent nor sensitive, yet it is not an inconsistency to suppose at the same time, that the animating power survives the body which it has animated, by laws as independent of any supernatural agent as those through which it first became united with it. Nor, if a future state be clearly proved, does it follow that it will be a state of punishment or reward.

By the word death, we express that condition in which natures resembling ourselves apparently cease to be that which they were. We no longer hear them speak, nor see them move. If they have sensations and apprehensions, we no longer participate in them. We know no more than that those external organs, and all that fine texture of material frame, without which we have no experience that life or thought can subsist, are dissolved and scattered abroad. The body is placed under the earth, and after a certain period there remains no vestige even of its form. This is that contemplation of inexhaustible melancholy, whose shadow eclipses the brightness of the world. The common observer is struck with dejection at the spectacle. He contends in vain against the persuasion of the grave, that the dead indeed cease to be. The corpse at his feet is prophetic of his own destiny. Those who have preceded him, and whose voice was delightful to his ear; whose touch met his like sweet and subtle fire; whose aspect spread a visionary light upon his path—these he cannot meet again. The organs of sense are destroyed, and the intellectual operations dependent on them have perished with their sources. How can a corpse see or feel? its eyes are eaten out, and its heart is black and without motion. What intercourse can two heaps of putrid clay and crumbling bones hold together? When you can discover where the fresh colours of the faded flower abide, or the music of the broken lyre, seek life among the dead. Such are the anxious and fearful contemplations of the common observer, though the popular religion often prevents him from confessing them even to himself.

The natural philosopher, in addition to the sensations common to all men inspired by the event of death, believes that he sees with more certainty that it is attended with the annihilation of sentiment and thought. He observes the mental powers increase and fade with those of the body, and even accommodate themselves to the most transitory changes of our physical nature. Sleep suspends many of the faculties of the vital and intellectual principle ; drunkenness and disease will either temporarily or permanently derange them. Madness or idiotcy may utterly extinguish the most excellent and delicate of those powers. In old age the mind gradually withers ; and as it grew and was strengthened with the body, so does it together with the body sink into decrepitude. Assuredly these are convincing evidences that so soon as the organs of the body are subjected to the laws of inanimate matter, sensation, and perception, and apprehension, are at an end. It is probable that what we call thought is not an actual being, but no more than the relation between certain parts of that infinitely varied mass, of which the rest of the universe is composed, and which ceases to exist so soon as those parts change their position with regard to each other. Thus colour, and sound, and taste, and odour exist only relatively. But let thought be considered as some peculiar substance, which permeates, and is the cause of, the animation of living beings. Why should that substance be assumed to be something essentially distinct from all others, and exempt from subjection to those laws from which no other substance is exempt ? It differs, indeed, from all other substances, as electricity, and light, and magnetism, and the constituent parts of air and earth, severally differ from all others. Each of these is subject to change and to decay, and to conversion into other forms. Yet the difference between light and earth is scarcely greater than that which exists between life, or thought, and fire. The difference between the two former was never alleged as an argument for the eternal permanence of either, in that form under which they first might offer themselves to our notice. Why should the difference between the two latter substances be an argument for the prolongation of the existence of one and not the other, when the existence of both has arrived at their apparent termination? To say that fire exists without manifesting any of the properties of fire, such as light, heat, &c., or that the principle of life exists without consciousness, or memory, or desire, or motive, is to resign, by an awkward distortion of language, the affirmative of the dispute. To say that the principle of life *may* exist in distribution among various forms, is to assert what cannot be proved to be either true or false, but which, were it true, annihilates all hope of existence after death, in any sense in which that event can belong to the hopes and fears of men. Suppose, however, that the intellectual and vital principle differs in the most marked and essential manner from all other known substances; that they have all some resemblance between themselves which it in no degree participates. In what manner can this concession be made an argument for its imperishability? All that we see or know perishes and is changed. Life and thought differ indeed from everything else. But that it survives that period, beyond which we have no experience of its existence, such distinction and dissimilarity affords no shadow of proof, and nothing but our own desires could have led us to conjecture or imagine.

Have we existed before birth? It is difficult to conceive the possibility of this. There is, in the generative principle of each animal and plant, a power which converts the substances by which it is surrounded into a substance homogeneous with itself. That is, the relations between certain elementary particles of matter undergo a change, and submit to new combinations. For when we use the words *principle*, *power*, *cause*, &c., we mean to express no real being, but only to class under those terms a certain series of co-existing phenomena ; but let it be supposed that this principle is a certain substance which escapes the observation of the chemist and anatomist. It certainly *may be* ; though it is sufficiently unphilosophical to allege the possibility of an opinion as a proof of its truth. Does it see, hear, feel, before its combination with those organs on which sensation depends ? Does it reason, imagine, apprehend, without those ideas which sensation alone can communicate ? If we have not existed before birth ; if, at the period when the parts of our nature on which thought and life depend, seem to be woven together, they are woven together ; if there are no reasons to suppose that we have existed before that period at which our existence apparently commences, then there are no grounds for supposition that we shall continue to exist after our existence has apparently ceased. So far as thought and life is concerned, the same will take place with regard to us, individually considered, after death, as had place before our birth.

It is said that it is possible that we should continue to exist in some mode totally inconceivable to us at present. This is a most unreasonable presumption. It casts on the adherents of annihilation the burthen of proving the negative of a question, the affirmative of which is not supported by a single argument, and which, by its very nature, lies beyond the experience of the human understanding. It is sufficiently easy, indeed, to form any proposition, concerning which we are ignorant, just not so absurd as not to be contradictory in itself, and defy refutation. The possibility of whatever enters into the wildest imagination to conceive is thus triumphantly vindicated. But it is enough that such assertions should be either contradictory to the known laws of nature, or exceed the limits of our experience, that their fallacy or irrelevancy to our consideration should be demonstrated. They persuade, indeed, only those who desire to be persuaded.

This desire to be for ever as we are; the reluctance to a violent and unexperienced change, which is common to all the animated and inanimate combinations of the universe, is, indeed, the secret persuasion which has given birth to the opinions of a future state.

1815

# SPECULATIONS ON METAPHYSICS

## I.—THE MIND

It is an axiom in mental philosophy, that we can think of nothing which we have not perceived. When I say that we can think of nothing, I mean, we can imagine nothing, we can reason of nothing, we can remember nothing, we can foresee nothing. The most astonishing combinations of poetry, the subtlest deductions of logic and mathematics, are no other than combinations which the intellect makes of sensations according to its own laws. A catalogue of all the thoughts of the mind, and of all their possible modifications, is a cyclopedic history of the universe.

But, it will be objected, the inhabitants of the various planets of this and other solar systems; and the existence of a Power bearing the same relation to all that we perceive and are, as what we call a cause does to what we call effect, were never subjects of sensation, and yet the laws of mind almost universally suggest, according to the various disposition of each, a conjecture, a persuasion, or a conviction of their existence. The reply is simple; these thoughts are also to be included in the catalogue of existence;

they are modes in which thoughts are combined; the objection only adds force to the conclusion, that beyond the limits of perception and thought nothing can exist.

Thoughts, or ideas, or notions, call them what you will, differ from each other, not in kind, but in force. It has commonly been supposed that those distinct thoughts which affect a number of persons, at regular intervals, during the passage of a multitude of other thoughts, which are called *real* or *external objects*, are totally different in kind from those which affect only a few persons, and which recur at irregular intervals, and are usually more obscure and indistinct, such as hallucinations, dreams, and the ideas of madness. No essential distinction between any one of these ideas, or any class of them, is founded on a correct observation of the nature of things, but merely on a consideration of what thoughts are most invariably subservient to the security and happiness of life; and if nothing more were expressed by the distinction, the philosopher might safely accommodate his language to that of the vulgar. But they pretend to assert an essential difference, which has no foundation in truth, and which suggests a narrow and false conception of universal nature, the parent of the most fatal errors in speculation. A specific difference between every thought of the mind, is, indeed, a necessary consequence of that law by which it perceives diversity and number; but a generic and essential difference is wholly arbitrary. The principle of the agreement and similarity of all thoughts, is, that they are all thoughts; the principle of their disagreement consists in the variety and irregularity of the occasions on which they arise in the mind. That in which they agree, to that in which they differ, is as everything to nothing. Important distinctions, of various degrees of force, indeed, are to be established between them, if they were, as they may be, subjects of ethical and economical discussion ; but that is a question altogether distinct.

By considering all knowledge as bounded by perception, whose operations may be indefinitely combined, we arrive at a conception of Nature inexpressibly more magnificent, simple and true, than accords with the ordinary systems of complicated and partial consideration. Nor does a contemplation of the universe, in this comprehensive and synthetical view, exclude the subtlest analysis of its modifications and parts.

A scale might be formed, graduated according to the degrees of a combined ratio of intensity, duration, connexion, periods of recurrence, and utility, which would be the standard, according to which all ideas might be measured, and an uninterrupted chain of nicely shadowed distinctions would be observed, from the faintest impression on the senses, to the most distinct combination of those impressions ; from the simplest of those combinations, to that mass of knowledge which, including our own nature, constitutes what we call the universe.

We are intuitively conscious of our own existence, and of that connexion in the train of our successive ideas, which we term our identity. We are conscious also of the existence of other minds ; but not intuitively. Our evidence, with respect to the existence of other minds, is founded upon a very complicated relation of ideas, which it is foreign to the purpose of this treatise to anatomize. The basis of this relation is, undoubtedly, a periodical recurrence of masses of ideas, which our voluntary determinations have, in one peculiar direction, no power to circumscribe or to arrest, and against the recurrence of which they can only imperfectly provide. The irresistible laws of thought constrain us to believe that the precise limits of our actual ideas are not the actual limits of possible ideas; the law, according to which these deductions are drawn, is called analogy; and this is the foundation of all our inferences, from one idea to another, inasmuch as they resemble each other.

We see trees, houses, fields, living beings in our own shape, and in shapes more or less analogous to our own. These are perpetually changing the mode of their existence relatively to us. To express the varieties of these modes, we say, *we move, they move*; and as this motion is continual, though not uniform, we express our conception of the diversities of its course by—*it has been, it is, it shall be.* These diversities are events or objects, and are essential, considered relatively to human identity, for the existence of the human mind. For if the inequalities, produced by what has been termed the operations of the external universe, were levelled by the perception of our being, uniting and filling up their interstices, motion and mensuration, and time, and space; the elements of the human mind being thus abstracted, sensation and imagination cease. Mind cannot be considered pure.

## II.—WHAT METAPHYSICS ARE. ERRORS IN THE USUAL METHODS OF CONSIDERING THEM

We do not attend sufficiently to what passes within ourselves. We combine words, combined a thousand times before. In our minds we assume entire opinions; and in the expression of those opinions, entire phrases, when we would philosophize. Our whole style of expression and sentiment is infected with the tritest plagiarisms. Our words are dead, our thoughts are cold and borrowed.

Let us contemplate facts ; let us, in the great study of ourselves, resolutely compel the mind to a rigid consideration of itself. We are not content with conjecture, and inductions, and syllogisms, in sciences regarding external objects. As in these, let us also, in considering the phenomena of mind, severely collect those facts which cannot be disputed. Metaphysics will thus possess this conspicuous advantage over every other science, that each student, by attentively referring to his own mind, may ascertain the authorities upon which any assertions regarding it are supported. There can thus be no deception, we ourselves being the depositaries of the evidence of the subject which we consider.

Metaphysics may be defined as an inquiry concerning those things belonging to, or connected with, the internal nature of man.

It is said that mind produces motion ; and it might as well have been said, that motion produces mind.

## III.—DIFFICULTY OF ANALYSING THE HUMAN MIND

If it were possible that a person should give a faithful history of his being, from the earliest epochs of his recollection, a picture would be presented such as the world has never contemplated before. A mirror would be held up to all men in which they might behold their own recollections, and, in dim perspective, their shadowy hopes and fears,—all that they dare not, or that, daring and desiring, they could not expose to the open eyes of day. But thought can with difficulty visit the intricate and winding chambers which it inhabits. It is like a river whose rapid and perpetual stream flows outwards ;—like one in dread who speeds through the recesses of some haunted pile, and dares not look behind. The caverns of the mind are obscure, and shadowy; or pervaded with a lustre, beautifully bright indeed, but shining not beyond their portals. If it were possible to be where we have been, vitally and indeed—if, at the moment of our presence there, we could define the results of our experience,—if the passage from sensation to reflection—from a state of passive perception to voluntary contemplation, were not so dizzying and so tumultuous, this attempt would be less difficult.

## IV.—HOW THE ANALYSIS SHOULD BE CARRIED ON

Most of the errors of philosophers have arisen from considering the human being in a point of view too detailed and circumscribed. He is not a moral, and an intellectual,—but also, and pre-eminently, an imaginative being. His own mind is his law; his own mind is all things to him. If we would arrive at any knowledge which should be serviceable from the practical conclusions to which it leads, we ought to consider the mind of man and the universe as the great whole on which to exercise our speculations. Here, above all, verbal disputes ought to be laid aside, though this has long been their chosen field of battle. It imports little to inquire whether thought be distinct from the objects of thought. The use of the words *external* and *internal*, as applied to the establishment of this distinction, has been the symbol and the source of much dispute. This is merely an affair of words, and as the dispute deserves, to say, that when speaking of the objects of thought, we indeed only describe one of the forms of thought—or that, speaking of thought, we only apprehend one of the operations of the universal system of beings.

## V.—CATALOGUE OF THE PHENOMENA OF DREAMS, AS CONNECTING SLEEPING AND WAKING

1. LET us reflect on our infancy, and give as faithfully as possible a relation of the events of sleep.

And first I am bound to present a faithful picture of my own peculiar nature relatively to sleep. I do not doubt that were every individual to imitate me, it would be found that among many circumstances peculiar to their individual nature, a sufficiently general resemblance would be found to prove the connexion existing between those peculiarities and the most universal phenomena. I shall employ caution, indeed, as to the facts which I state, that they contain nothing false or exaggerated. But they contain no more than certain elucidations of my own nature; concerning the degree in which it resembles, or differs from, that of others, I am by no means accurately aware. It is sufficient, however, to caution the reader against drawing general inferences from particular instances.

I omit the general instances of delusion in fever or delirium, as well as mere dreams considered in themselves. A delineation of this subject, however inexhaustible and interesting, is to be passed over.

What is the connexion of sleeping and of waking?

2. I distinctly remember dreaming three several times, between intervals of two or more years, the same precise dream. It was not so much what is ordinarily called a dream; the single image, unconnected with all other images, of a youth who was educated at the same school with myself, presented itself in sleep. Even now, after the lapse of many years, I can never hear the name of this youth, with-

out the three places where I dreamed of him presenting themselves distinctly to my mind.

3. In dreams, images acquire associations peculiar to dreaming; so that the idea of a particular house, when it recurs a second time in dreams, will have relation with the idea of the same house, in the first time, of a nature entirely different from that which the house excites, when seen or thought of in relation to waking ideas.

4. I have beheld scenes, with the intimate and unaccountable connexion of which with the obscure parts of my own nature, I have been irresistibly impressed. I have beheld a scene which has produced no unusual effect on my thoughts. After the lapse of many years I have dreamed of this scene. It has hung on my memory, it has haunted my thoughts, at intervals, with the pertinacity of an object connected with human affections. I have visited this scene again. Neither the dream could be dissociated from the landscape, nor the landscape from the dream, nor feelings, such as neither singly could have awakened, from both. But the most remarkable event of this nature, which ever occurred to me, happened five years ago at Oxford. I was walking with a friend, in the neighbourhood of that city, engaged in earnest and interesting conversation. We suddenly turned the corner of a lane, and the view, which its high banks and hedges had concealed, presented itself. The view consisted of a windmill, standing in one among many plashy meadows, inclosed with stone walls; the irregular and broken ground, between the wall and the road on which we stood; a long low hill behind the windmill, and a grey covering of uniform cloud spread over the evening sky. It was that season when the last leaf had just fallen from the scant and stunted ash. The scene surely was a common

scene; the season and the hour little calculated to kindle lawless thought; it was a tame uninteresting assemblage of objects, such as would drive the imagination for refuge in serious and sober talk, to the evening fireside, and the dessert of winter fruits and wine. The effect which it produced on me was not such as could have been expected. I suddenly remembered to have seen that exact scene in some dream of long[1]———

1815

# SPECULATIONS ON MORALS

## I.—PLAN OF A TREATISE ON MORALS

THAT great science which regards nature and the operations of the human mind, is popularly divided into Morals and Metaphysics. The latter relates to a just classification, and the assignment of distinct names to its ideas; the former regards simply the determination of that arrangement of them which produces the greatest and most solid happiness. It is admitted that a virtuous or moral action, is that action which, when considered in all its accessories and consequences, is fitted to produce the highest pleasure to the greatest number of sensitive beings. The laws according to which all pleasure, since it cannot be equally felt by all sensitive beings, ought to be distributed by a voluntary agent, are reserved for a separate chapter.

The design of this little treatise is restricted to the development of the elementary principles of morals. As far as regards that purpose, metaphysical science will be treated merely so far as a source of negative truth; whilst morality will be considered as a

[1] Here I was obliged to leave off, overcome by thrilling horror.

science, respecting which we can arrive at positive conclusions.

The misguided imaginations of men have rendered the ascertaining of what *is not true*, the principal direct service which metaphysical science can bestow upon moral science. Moral science itself is the doctrine of the voluntary actions of man, as a sentient and social being. These actions depend on the thoughts in his mind. But there is a mass of popular opinion, from which the most enlightened persons are seldom wholly free, into the truth or falsehood of which it is incumbent on us to inquire, before we can arrive at any firm conclusions as to the conduct which we ought to pursue in the regulation of our own minds, or towards our fellow beings; or before we can ascertain the elementary laws, according to which these thoughts, from which these actions flow, are originally combined.

The object of the forms according to which human society is administered, is the happiness of the individuals composing the communities which they regard, and these forms are perfect or imperfect in proportion to the degree in which they promote this end.

This object is not merely the quantity of happiness enjoyed by individuals as sensitive beings, but the mode in which it should be distributed among them as social beings. It is not enough, if such a coincidence can be conceived as possible, that one person or class of persons should enjoy the highest happiness, whilst another is suffering a disproportionate degree of misery. It is necessary that the happiness produced by the common efforts, and preserved by the common care, should be distributed according to the just claims of each individual; if not, although the quantity produced should be the same, the end of society would remain unfulfilled. The object is in a compound proportion to the quantity of happiness produced, and the correspondence of the mode in which it is distributed, to the elementary feelings of man as a social being.

The disposition in an individual to promote this object is called virtue; and the two constituent parts of virtue, benevolence and justice, are correlative with these two great portions of the only true object of all voluntary actions of a human being. Benevolence is the desire to be the author of good, and justice the apprehension of the manner in which good ought to be done.

Justice and benevolence result from the elementary laws of the human mind.

## CHAPTER I

### ON THE NATURE OF VIRTUE

SECT. 1. General View of the Nature and Objects of Virtue.—2. The Origin and Basis of Virtue, as founded on the Elementary Principles of Mind.—3. The Laws which flow from the nature of Mind regulating the application of those principles to human actions.—4. Virtue, a possible attribute of man.

WE exist in the midst of a multitude of beings like ourselves, upon whose happiness most of our actions exert some obvious and decisive influence.

The regulation of this influence is the object of moral science.

We know that we are susceptible of receiving painful or pleasurable impressions of greater or less intensity and duration. That is called good which produces pleasure; that is called evil which produces pain. These are general names, applicable to every class of causes, from which an overbalance of pain or pleasure may result. But when a human being is the active instrument of generating or diffusing happiness, the principle through which it is most effectually instrumental to that purpose, is called virtue. And benevolence, or the desire to be the author of good, united with justice, or an apprehension of the manner in which that good is to be done, constitutes virtue.

But wherefore should a man be benevolent and just? The immediate emotions of his nature, especially in its most inartificial state, prompt him to inflict pain, and to arrogate dominion. He desires to heap superfluities to his own store, although others perish with famine. He is propelled to guard against the smallest invasion of his own liberty, though he reduces others to a condition of the most pitiless servitude. He is revengeful, proud and selfish. Wherefore should he curb these propensities?

It is inquired, for what reason a human being should engage in procuring the happiness, or refrain from producing the pain of another? When a reason is required to prove the necessity of adopting any system of conduct, what is it that the objector demands? He requires proof of that system of conduct being such as will most effectually promote the happiness of mankind. To demonstrate this, is to render a moral reason. Such is the object of Virtue.

A common sophism, which, like many others, depends on the abuse of a metaphorical expression to a literal purpose, has produced much of the confusion which has involved the theory of morals. It is said that no person is bound to be just or kind, if, on his neglect, he should fail to incur some penalty. Duty is obligation. There can be no obligation without an obliger. Virtue is a law, to which it is the will of the lawgiver that we should conform; which will we should in no manner be bound to obey, unless some dreadful punishment were attached to disobedience. This is the philosophy of slavery and superstition.

In fact, no person can be *bound* or *obliged*, without some power preceding to bind and oblige. If I observe a man bound hand and foot, I know that some one bound him. But if I observe him returning self-satisfied from the performance of some action, by which he has been the willing author of extensive benefit, I do not infer that the anticipation of hellish agonies, or the hope of heavenly reward, has constrained him to such an act.

. . . . . . . . .

It remains to be stated in what manner the sensations which constitute the basis of virtue originate in the human mind; what are the laws which it receives there; how far the principles of mind allow it to be an attribute of a human being; and, lastly, what is the probability of persuading mankind to adopt it as a universal and systematic motive of conduct.

### BENEVOLENCE

There is a class of emotions which we instinctively avoid. A human being, such as is man considered in his origin, a child a month old, has a very imperfect consciousness of the existence of other natures resembling itself. All the energies of its being are directed to the extinction of the pains with which it is perpetually assailed. At length it discovers that it is surrounded by natures susceptible of sensations similar to its own. It is very late before children attain to this knowledge. If a child observes, without emotion, its nurse or its mother suffering acute pain, it is attributable rather to ignorance than insensibility. So soon as the accents and gestures, significant of pain, are referred to the feelings which they express, they awaken in the mind of the beholder a desire that they should cease. Pain is thus apprehended to be evil for its own sake, without any other necessary reference to the mind by which its existence is perceived, than such as is indispensable to its perception. The tendencies of our original sensations, indeed, all have for their object the preservation of our individual being. But these are passive and unconscious. In proportion as the mind acquires an active power, the empire of these tendencies becomes limited. Thus an infant, a savage, and a solitary beast, is selfish, because its mind is incapable of receiving an accurate intimation of the nature of pain as existing in beings resembling itself. The inhabitant of a highly civilized community will more acutely sympathize with the sufferings and enjoyments of others, than the inhabitant of a society of a less degree of civilization. He who shall have cultivated his intellectual powers by familiarity with the highest specimens of poetry and philosophy, will usually sympathize more than one engaged in the less refined functions of manual labour. Every one has experience of the fact, that to sympathize with the sufferings of another, is to enjoy a transitory oblivion of his own.

The mind thus acquires, by exercise, a habit, as it were, of perceiving and abhorring evil, however remote from the immediate sphere of sensations with which that individual mind is conversant. Imagination or mind employed in prophetically imaging forth its objects, is that faculty of human nature on which every gradation of its progress, nay, every, the minutest, change, depends. Pain or pleasure, if subtly analysed, will be found to consist entirely in prospect. The only distinction between the selfish man and the virtuous man is, that the imagination of the former is confined within a narrow limit, whilst that of the latter embraces a comprehensive circumference. In this sense, wisdom and virtue may be said to be inseparable, and criteria of each other. Selfishness is the offspring of ignorance and mistake; it is the portion of unreflecting infancy, and savage solitude, or of those whom toil or evil occupations have blunted or rendered torpid; disinterested benevolence is the product of a cultivated imagination, and has an intimate connexion with all the arts which add ornament, or dignity, or power, or stability to the social state of man. Virtue is thus entirely a refinement of civilized life; a creation of the human mind; or, rather, a combination which it has made, according to elementary rules contained within itself, of the feelings suggested by the relations established between man and man.

All the theories which have refined and exalted humanity, or those which have been devised as alleviations of its mistakes and evils, have been based upon the elementary emotions of disinterestedness, which we feel to constitute the majesty of our nature. Patriotism, as it existed in the ancient republics, was never, as has been supposed, a calculation of personal advantages. When Mutius Scaevola thrust his hand into the burning coals, and Regulus returned to Carthage, and Epicharis sustained the rack silently, in the torments of which she knew that she would speedily perish, rather than betray the conspirators to the tyrant;[1] these illustrious persons certainly made a small estimate of their private interest. If it be said that they sought posthumous fame; instances are not wanting in history which prove that men have even defied infamy for the sake of good. But there is a great error in the world with respect to the selfishness

[1] Tacitus.

of fame. It is certainly possible that a person should seek distinction as a medium of personal gratification. But the love of fame is frequently no more than a desire that the feelings of others should confirm, illustrate, and sympathize with, our own. In this respect it is allied with all that draws us out of ourselves. It is the 'last infirmity of noble minds'. Chivalry was likewise founded on the theory of self-sacrifice. Love possesses so extraordinary a power over the human heart, only because disinterestedness is united with the natural propensities. These propensities themselves are comparatively impotent in cases where the imagination of pleasure to be given, as well as to be received, does not enter into the account. Let it not be objected that patriotism, and chivalry, and sentimental love, have been the fountains of enormous mischief. They are cited only to establish the proposition that, according to the elementary principles of mind, man is capable of desiring and pursuing good for its own sake.

### JUSTICE

The benevolent propensities are thus inherent in the human mind. We are impelled to seek the happiness of others. We experience a satisfaction in being the authors of that happiness. Everything that lives is open to impressions of pleasure and pain. We are led by our benevolent propensities to regard every human being indifferently with whom we come in contact. They have preference only with respect to those who offer themselves most obviously to our notice. Human beings are indiscriminating and blind; they will avoid inflicting pain, though that pain should be attended with eventual benefit; they will seek to confer pleasure without calculating the mischief that may result. They benefit one at the expense of many.

There is a sentiment in the human mind that regulates benevolence in its application as a principle of action. This is the sense of justice. Justice, as well as benevolence, is an elementary law of human nature. It is through this principle that men are impelled to distribute any means of pleasure which benevolence may suggest the communication of to others, in equal portions among an equal number of applicants. If ten men are shipwrecked on a desert island, they distribute whatever subsistence may remain to them, into equal portions among themselves. If six of them conspire to deprive the remaining four of their share, their conduct is termed unjust.

The existence of pain has been shown to be a circumstance which the human mind regards with dissatisfaction, and of which it desires the cessation. It is equally according to its nature to desire that the advantages to be enjoyed by a limited number of persons should be enjoyed equally by all. This proposition is supported by the evidence of indisputable facts. Tell some ungarbled tale of a number of persons being made the victims of the enjoyments of one, and he who would appeal in favour of any system which might produce such an evil to the primary emotions of our nature, would have nothing to reply. Let two persons, equally strangers, make application for some benefit in the possession of a third to bestow, and to which he feels that they have an equal claim. They are both sensitive beings; pleasure and pain affect them alike.

. . . . . . . .

## CHAPTER II

It is foreign to the general scope of this little treatise to encumber a simple argument by controverting any of the trite objections of habit or fanaticism. But there are two; the first, the basis of all political mistake, and the second, the prolific cause and effect of religious error, which it seems useful to refute.

First, it is inquired, 'Wherefore should a man be benevolent and just?' The answer has been given in the preceding chapter.

If a man persists to inquire why he ought to promote the happiness of mankind, he demands a mathematical or metaphysical reason for a moral action. The absurdity of this scepticism is more apparent, but not less real than the exacting a moral reason for a mathematical or metaphysical fact. If any person should refuse to admit that all the radii of a circle are of equal length, or that human actions are necessarily determined by motives, until it could be proved that these radii and these actions uniformly tended to the production of the greatest general good, who would not wonder at the unreasonable and capricious association of his ideas?

The writer of a philosophical treatise may, I imagine, at this advanced era of human intellect, be held excused from entering into a controversy with those reasoners, if such there are, who would claim an exemption from its decrees in favour of any one among those diversified systems of obscure opinion respecting morals, which, under the name of religions, have in various ages and countries prevailed among mankind. Besides that if, as these reasoners have pretended, eternal torture or happiness will ensue as the consequence of certain actions, we should be no nearer the possession of a standard to determine what actions were right and wrong, even if this pretended revelation, which is by no means the case, had furnished us with a complete catalogue of them. The character of actions as virtuous or vicious would by no means be determined alone by the personal advantage or disadvantage of each moral agent individually considered. Indeed, an action is often virtuous in proportion to the greatness of the personal calamity which the author willingly draws upon himself by daring to perform it. It is because an action produces an overbalance of pleasure or pain to the greatest number of sentient beings, and not merely because its consequences are beneficial or injurious to the author of that action, that it is good or evil. Nay, this latter consideration has a tendency to pollute the purity of virtue, inasmuch as it consists in the motive rather than in the consequences of an action. A person who should labour for the happiness of mankind lest he should be tormented eternally in Hell, would, with reference to that motive, possess as little claim to the epithet of virtuous, as he who should torture, imprison, and burn them alive, a more usual and natural consequence of such principles, for the sake of the enjoyments of Heaven.

My neighbour, presuming on his strength, may direct me to perform or to refrain from a particular action; indicating a certain arbitrary penalty in the event of disobedience within his power to inflict. My action, if modified by his menaces, can in no degree participate in virtue. He has afforded me no criterion as to what is right or wrong. A king, or an assembly of men, may publish a proclamation affixing any penalty to any particular action, but that is not immoral because such penalty is affixed. Nothing is more evident than that the epithet of virtue is inapplicable to the refraining from that action on account of the evil arbitrarily attached to it. If the action is in itself beneficial, virtue would rather consist in not refraining from it, but in firmly defying the personal consequences attached to its performance.

Some usurper of supernatural energy might subdue the whole globe to his power ; he might possess new and unheard-of resources for enduing his punishments with the most terrible attributes of pain. The torments of his victims might be intense in their degree, and protracted to an infinite duration. Still the 'will of the lawgiver' would afford no surer criterion as to what actions were right or wrong. It would only increase the possible virtue of those who refuse to become the instruments of his tyranny.

## II.—MORAL SCIENCE CONSISTS IN CONSIDERING THE DIFFERENCE, NOT THE RESEMBLANCE, OF PERSONS

The internal influence, derived from the constitution of the mind from which they flow, produces that peculiar modification of actions, which makes them intrinsically good or evil.

To attain an apprehension of the importance of this distinction, let us visit, in imagination, the proceedings of some metropolis. Consider the multitude of human beings who inhabit it, and survey, in thought, the actions of the several classes into which they are divided. Their obvious actions are apparently uniform : the stability of human society seems to be maintained sufficiently by the uniformity of the conduct of its members, both with regard to themselves, and with regard to others. The labourer arises at a certain hour, and applies himself to the task enjoined him. The functionaries of government and law are regularly employed in their offices and courts. The trader holds a train of conduct from which he never deviates. The ministers of religion employ an accustomed language, and maintain a decent and equable regard. The army is drawn forth, the motions of every soldier are such as they were expected to be; the general commands, and his words are echoed from troop to troop. The domestic actions of men are, for the most part, undistinguishable one from the other, at a superficial glance. The actions which are classed under the general appellation of marriage, education, friendship, &c., are perpetually going on, and to a superficial glance, are similar one to the other.

But, if we would see the truth of things, they must be stripped of this fallacious appearance of uniformity. In truth, no one action has, when considered in its whole extent, any essential resemblance with any other. Each individual, who composes the vast multitude which we have been contemplating, has a peculiar frame of mind, which, whilst the features of the great mass of his actions remain uniform, impresses the minuter lineaments with its peculiar hues. Thus, whilst his life, as a whole, is like the lives of other men, in detail, it is most unlike; and the more subdivided the actions become; that is, the more they enter into that class which have a vital influence on the happiness of others and his own, so much the more are they distinct from those of other men.

> Those little, nameless, unremembered acts
> Of kindness and of love,

as well as those deadly outrages which are inflicted by a look, a word—or less—the very refraining from some faint and most evanescent expression of countenance; these flow from a profounder source than the series of our habitual conduct, which, it has been already said, derives its origin from without. These are the actions, and such as these, which make human life what it is, and are the fountains of all the good and evil with which its entire surface is so widely and impartially overspread; and though they are called minute, they are called so in compliance with the blindness of those who cannot estimate their importance. It is in the due appreciating the general effects of their peculiarities, and in cultivating the habit of acquiring decisive knowledge respecting the tendencies arising out of them in particular cases, that the most important part of moral science consists. The deepest abyss of these vast and multitudinous caverns, it is necessary that we should visit.

This is the difference between social and individual man. Not that this distinction is to be considered definite, or characteristic of one human being as compared with another; it denotes rather two classes of agency, common in a degree to every human being. None is exempt, indeed, from that species of influence which affects, as it were, the surface of his being, and gives the specific outline to his conduct. Almost all that is ostensible submits to that legislature created by the general representation of the past feelings of mankind—imperfect as it is from a variety of causes, as it exists in the government, the religion, and domestic habits. Those who do not nominally, yet actually, submit to the same power. The external features of their conduct, indeed, can no more escape it, than the clouds can escape from the stream of the wind; and his opinion, which he often hopes he has dispassionately secured from all contagion of prejudice and vulgarity, would be found, on examination, to be the inevitable excrescence of the very usages from which he vehemently dissents. Internally all is conducted otherwise; the efficiency, the essence, the vitality of actions, derives its colour from what is no ways contributed to from any external source. Like the plant, which while it derives the accident of its size and shape from the soil in which it springs, and is cankered, or distorted, or inflated, yet retains those qualities which essentially divide it from all others; so that hemlock continues to be poison, and the violet does not cease to emit its odour in whatever soil it may grow.

We consider our own nature too superficially. We look on all that in ourselves with which we can discover a resemblance in others; and consider those resemblances as the materials of moral knowledge. It is in the differences that it actually consists. 1815

# ESSAY ON CHRISTIANITY

THE Being who has influenced in the most memorable manner the opinions and the fortunes of the human species, is Jesus Christ. At this day, his name is connected with the devotional feelings of two hundred millions of the race of man. The institutions of the most civilized portion of the globe derive their authority from the sanction of his doctrines; he is the hero, the God, of our popular religion. His extraordinary genius, the wide and rapid effect of his unexampled doctrines, his invincible gentleness and benignity, the devoted love borne to him by his adherents, suggested a persuasion to them that he was something divine. The supernatural events which the historians of this wonderful man subsequently asserted to have been connected with every gradation of his career, established the opinion.

His death is said to have been accompanied by an accumulation of tremendous prodigies. Utter darkness fell upon the earth, blotting the noonday sun; dead bodies, arising from their graves, walked through the public streets, and an earthquake shook the astonished city, rending the rocks of the surrounding mountains. The philosopher may attribute the application of these events to the death of a reformer, or the events themselves to a visitation of that universal Pan who——

. . . . . .

The thoughts which the word 'God' suggests to the human mind are susceptible of as many variations as human minds themselves. The Stoic, the Platonist, and the Epicurean, the Polytheist, the Dualist, and the Trinitarian, differ infinitely in their conceptions of its meaning. They agree only in considering it the most awful and most venerable of names, as a common term devised to express all of mystery, or majesty, or power, which the invisible world contains. And not only has every sect distinct conceptions of the application of this name, but scarcely two individuals of the same sect, who exercise in any degree the freedom of their judgement, or yield themselves with any candour of feeling to the influences of the visible world, find perfect coincidence of opinion to exist between them. It is [interesting] to inquire in what acceptation Jesus Christ employed this term.

We may conceive his mind to have been predisposed on this subject to adopt the opinions of his countrymen. Every human being is indebted for a multitude of his sentiments to the religion of his early years. Jesus Christ probably [studied] the historians of his country with the ardour of a spirit seeking after truth. They were undoubtedly the companions of his childish years, the food and nutriment and materials of his youthful meditations. The sublime dramatic poem entitled *Job* had familiarized his imagination with the boldest imagery afforded by the human mind and the material world. *Ecclesiastes* had diffused a seriousness and solemnity over the frame of his spirit, glowing with youthful hope, and [had] made audible to his listening heart

> The still, sad music of humanity,
> Not harsh or grating, but of ample power
> To chasten and subdue.

He had contemplated this name as having been profanely perverted to the sanctioning of the most enormous and abominable crimes. We can distinctly trace, in the tissue of his doctrines, the persuasion that God is some universal Being, differing from man and the mind of man. According to Jesus Christ, God is neither the Jupiter, who sends rain upon the earth; nor the Venus, through whom all living things are produced; nor the Vulcan, who presides over the terrestrial element of fire; nor the Vesta, that preserves the light which is enshrined in the sun and moon and stars. He is neither the Proteus nor the Pan of the material world. But the word God, according to the acceptation of Jesus Christ, unites all the attributes which these denominations contain, and is the [interpoint] and overruling Spirit of all the energy and wisdom included within the circle of existing things. It is important to observe that the author of the Christian system had a conception widely differing from the gross imaginations of the vulgar relatively to the ruling Power of the universe. He everywhere represents this Power as something mysteriously and illimitably prevading the frame of things. Nor do his doctrines practically assume any proposition which they theoretically deny. They do not represent God as a limitless and inconceivable mystery; affirming, at the same time, his existence as a Being subject to passion and capable——

. . . . . .

'Blessed are the pure in heart, for they shall see God.' Blessed are those who have preserved internal sanctity of soul; who are conscious of no secret deceit; who are the same in act as they are in desire; who conceal no thought, no tendencies of thought, from their own conscience; who are faithful and sincere witnesses, before the tribunal of their own judgements, of all that passes within their mind. Such as these shall see God. What! after death, shall their awakened eyes behold the King of Heaven? Shall they stand in awe before the golden throne on which He sits, and gaze upon the venerable countenance of the paternal Monarch? Is this the reward of the virtuous and the pure? These are the idle dreams of the visionary, or the pernicious representations of impostors, who have fabricated from the very materials of wisdom a cloak for their own dwarfish or imbecile conceptions.

Jesus Christ has said no more than the most excellent philosophers have felt and expressed—that virtue is its own reward. It is true that such an expression as he has used was prompted by the energy of genius, and was the overflowing enthusiasm of a poet; but it is not the less literally true [because] clearly repugnant to the mistaken conceptions of the multitude. God, it has been asserted, was contemplated by Jesus Christ as every poet and every philosopher must have contemplated that mysterious principle. He considered that venerable word to express the overruling Spirit of the collective energy of the moral and material world. He affirms, therefore, no more than that a simple, sincere mind is the indispensable requisite of true science and true happiness. He affirms that a being of pure and gentle habits will not fail, in every thought, in every object of every thought, to be aware of benignant visitings from the invisible energies by which he is surrounded.

Whosoever is free from the contamination of luxury and licence, may go forth to the fields and to the woods, inhaling joyous renovation from the breath of Spring, or catching from the odours and sounds of Autumn some diviner mood of sweetest sadness, which improves the softened heart. Whosoever is no deceiver or destroyer of his fellow men—no liar, no flatterer, no murderer—may walk among his species, deriving, from the communion with all which they contain of beautiful or of majestic, some intercourse with the Universal God. Whosoever has maintained with his own heart the strictest correspondence of confidence, who dares to examine and to estimate every imagination which suggests itself to his mind—whosoever is that which he designs to become, and only aspires to that which the divinity of his own nature shall consider and approve—he has already seen God.

We live and move and think; but we are not the creators of our own origin and existence. We are not the arbiters of every motion of our own complicated nature; we are not the masters of our own imaginations and moods of mental being. There is a Power by which we are surrounded, like the atmosphere in which some motionless lyre is suspended, which visits with its breath our silent chords at will.

Our most imperial and stupendous qualities—those on which the majesty and the power of humanity is erected—are, relatively to the inferior portion of its mechanism, active and imperial; but they are the passive slaves of some higher and more omnipotent Power. This Power is God; and those who have seen God have, in the period of their purer and more perfect nature, been harmonized by their own will to so exquisite [a] consentaneity of power as to give forth divinest melody, when the breath of universal being sweeps over their frame. That those who are pure in heart shall see God, and that virtue is its own reward, may be considered as equivalent assertions. The former of these propositions is a metaphorical repetition of the latter. The advocates of literal interpretation have been the most efficacious enemies of those doctrines whose nature they profess to venerate. Thucydides, in particular, affords a number of instances calculated——

. . . . .

Tacitus says, that the Jews held God to be something eternal and supreme, neither subject to change nor to decay; therefore, they permit no statues in their cities or their temples. The universal Being can only be described or defined by negatives which deny his subjection to the laws of all inferior existences. Where indefiniteness ends, idolatry and anthropomorphism begin. God is, as Lucan has expressed,

> Quocunque vides, quodcunque moveris,
> Et caelum et virtus.

The doctrine of what some fanatics have termed 'a peculiar Providence'—that is, of some power beyond and superior to that which ordinarily guides the operations of the Universe, interfering to punish the vicious and reward the virtuous—is explicitly denied by Jesus Christ. The absurd and execrable doctrine of vengeance, in *all its shapes*, seems to have been contemplated by this great moralist with the profoundest disapprobation; nor would he permit the most venerable of names to be perverted into a sanction for the meanest and most contemptible propensities incident to the nature of man. 'Love your enemies, bless those who curse you, that ye may be the sons of your Heavenly Father, who makes the sun to shine on the good and on the evil, and the rain to fall on the just and unjust.' How monstrous a calumny have not impostors dared to advance against the mild and gentle author of this just sentiment, and against the whole tenor of his doctrines and his life, overflowing with benevolence and forbearance and compassion! They have represented him asserting that the Omnipotent God—that merciful and benignant Power who scatters equally upon the beautiful earth all the elements of security and happiness—whose influences are distributed to all whose natures admit of a participation in them—who sends to the weak and vicious creatures of his will all the benefits which they are capable of sharing—that this God has devised a scheme whereby the body shall live after its apparent dissolution, and be rendered capable of indefinite torture. He is said to have compared the agonies which the vicious shall then endure to the excruciations of a living body bound among the flames, and being consumed sinew by sinew, and bone by bone.

And this is to be done, not because it is supposed (and the supposition would be sufficiently detestable) that the moral nature of the sufferer would be improved by his tortures—it is done because it *is just* to be done. My neighbour, or my servant, or my child, has done me an injury, and it is just that he should suffer an injury in return. Such is the doctrine which Jesus Christ summoned his whole resources of persuasion to oppose. 'Love your enemy, bless those who curse you': such, he says, is the practice of God, and such must ye imitate if ye would be the children of God.

Jesus Christ would hardly have cited, as an example of all that is gentle and beneficent and compassionate, a Being who shall deliberately scheme to inflict on a large portion of the human race tortures indescribably intense and indefinitely protracted: who shall inflict them, too, without any mistake as to the true nature of pain—without any view to future good—merely because it is just.

This, and no other, is justice:—to consider, under all the circumstances and consequences of a particular case, how the greatest quantity and purest quality of happiness will ensue from any action; [this] is to be just, and there is no other justice. The distinction between justice and mercy was first imagined in the courts of tyrants. Mankind receive every relaxation of their tyranny as a circumstance of grace or favour.

Such was the clemency of Julius Caesar, who, having achieved by a series of treachery and bloodshed the ruin of the liberties of his country, receives the fame of mercy because, possessing the power to slay the noblest men of Rome, he restrained his sanguinary soul, arrogating to himself as a merit an abstinence from actions which if he had committed, he would only have added one other atrocity to his deeds. His assassins understood justice better. They saw the most virtuous and civilized community of mankind under the insolent dominion of one wicked man; and they murdered him. They destroyed the usurper of the liberties of their countrymen, not because they hated him, not because they would revenge the wrongs which they had sustained (Brutus, it is said, was his most familiar friend; most of the conspirators were habituated to domestic intercourse with the man whom they destroyed): it was in affection, inextinguishable love for all that is venerable and dear to the human heart, in the names of Country, Liberty, and Virtue; it was in a serious and solemn and reluctant mood, that these holy patriots murdered their father and their friend. They would have spared his violent death, if he could have deposited the rights which he had assumed. His own selfish and narrow nature necessitated the sacrifices they made. They required that he should change all those habits which debauchery and bloodshed had twined around the fibres of his inmost frame of thought; that he should participate with them and with his country those privileges which, having corrupted by assuming to himself, he would no longer value. They would have sacrificed their lives if they could have made him worthy of the sacrifice. Such are the feelings which Jesus Christ asserts to belong to the ruling Power of the world. He desireth not the death of a sinner: he makes the sun to shine upon the just and unjust.

The nature of a narrow and malevolent spirit is so essentially incompatible with happiness as to render it inaccessible to the influences of the benignant God. All that his own perverse propensities will permit him to receive, that God abundantly pours forth upon him. If there is the slightest overbalance of happiness, which can be allotted to the most atrocious offender, consistently with the nature of things, that is rigidly made his portion by the ever-watchful Power of God. In every case, the human mind enjoys the utmost pleasure which it is capable of enjoying. God is represented by Jesus Christ as the Power from which, and through which, the streams of all that is excellent and delightful flow; the Power which models, as they pass, all the elements of this mixed universe to the purest and most perfect shape which it belongs to their nature to assume. Jesus Christ attributes to this Power the faculty of Will. How far such a doctrine, in its ordinary sense, may be philosophically true, or how far Jesus Christ intentionally availed himself of a metaphor easily understood, is foreign to the subject to consider. This much is certain, that Jesus Christ represents God as the fountain of all goodness, the eternal enemy of pain and evil, the uniform and unchanging motive of the salutary operations of the material world. The supposition that this cause is excited to action by some principle analogous to the human will, adds weight to the persuasion that it is foreign to its beneficent nature to inflict the slightest pain. According to Jesus Christ, and according to the indisputable facts of the case, some evil spirit has dominion in this imperfect world. But there will come a time when the human mind shall be visited exclusively by the influences of the benignant Power. Men shall die, and their bodies shall rot under the ground; all the organs through which their knowledge and their feelings have flowed, or in which they have originated, shall assume other forms, and become ministrant to purposes the most foreign from their former tendencies. There is a time when we shall neither be heard or be seen by the multitude of beings like ourselves by whom we have been so long surrounded. They shall go to graves; where then?

It appears that we moulder to a heap of senseless dust; to a few worms, that arise and perish, like ourselves. Jesus Christ asserts that these appearances are fallacious, and that a gloomy and cold imagination alone suggests the conception that thought can cease to be. Another and a more extensive state of being, rather than the complete extinction of being, will follow from that mysterious change which we call Death. There shall be no misery, no pain, no fear. The empire of evil spirits extends not beyond the boundaries of the grave. The unobscured irradiations from the fountain-fire of all goodness shall reveal all that is mysterious and unintelligible, until the mutual communications of knowledge and of happiness throughout all thinking natures constitute a harmony of good that ever varies and never ends.

This is Heaven, when pain and evil cease, and when the Benignant Principle, untrammelled and uncontrolled, visits in the fullness of its power the universal frame of things. Human life, with all its unreal ills and transitory hopes, is as a dream, which departs before the dawn, leaving no trace of its evanescent hues. All that it contains of pure or of divine visits the passive mind in some serenest mood. Most holy are the feelings through which our fellow beings are rendered dear and [venerable] to the heart. The remembrance of their sweetness, and the completion of the hopes which they [excite], constitute, when we awaken from the sleep of life, the fulfilment of the prophecies of its most majestic and beautiful visions.

We die, says Jesus Christ; and, when we awaken from the languor of disease, the glories and the happiness of Paradise are around us. All evil and pain have ceased for ever. Our happiness also corresponds with, and is adapted to, the nature of what is most excellent in our being. We see God, and we see that he is good. How delightful a picture, even if it be not true! How magnificent is the conception which this bold theory suggests to the contemplation, even if it be no more than the imagination of some sublimest and most holy poet, who, impressed with the loveliness and majesty of his own nature, is impatient and discontented with the narrow limits which this imperfect life and the dark grave have assigned for ever as his melancholy portion. It is not to be believed that Hell, or punishment, was the conception of this daring mind. It is not to be believed that the most prominent group of this picture, which is framed so heart-moving and lovely—the accomplishment of all human hope, the extinction of all morbid fear and anguish—would consist of millions of sensitive beings enduring, in every variety of torture which Omniscient vengeance could invent, immortal agony.

Jesus Christ opposed with earnest eloquence the panic fears and hateful superstitions which have enslaved mankind for ages. Nations had risen against nations, employing the subtlest devices of mechanism and mind to waste, and excruciate, and overthrow. The great community of mankind had been subdivided into ten thousand communities, each organized for the ruin of the other. Wheel within wheel, the vast machine was instinct with the restless spirit of desolation. Pain had been inflicted; therefore, pain should be inflicted in return. Retaliation of injuries is the only remedy which can be applied to violence, because it teaches the injurer the true nature of his own conduct, and operates as a warning against its repetition. Nor must the same measure of calamity be returned as was received. If a man borrows a certain sum from me, he is bound to repay that sum. Shall no more be required of the enemy who destroys my reputation, or ravages my fields? It is just that he should suffer ten times the loss which he has inflicted, that the legitimate consequences of his deed may never be obliterated from his remembrance, and that others may clearly discern and feel the danger of invading the peace of human society. Such reasonings, and the impetuous feelings arising from them, have armed nation against nation, family against family, man against man.

An Athenian soldier, in the Ionian army which had assembled for the purpose of vindicating the liberty of the Asiatic Greeks, accidentally set fire to Sardis. The city, being composed of combustible materials, was burned to the ground. The Persians believed that this circumstance of aggression made it their duty to retaliate on Athens. They assembled successive expeditions on the most extensive scale. Every nation of the East was united to ruin the Grecian States. Athens was burned to the ground, the whole territory laid waste, and every living thing which it contained [destroyed]. After suffering and inflicting incalculable mischiefs, they desisted from their purpose only when they became impotent to effect it. The desire of revenge for the aggression of Persia outlived, among the Greeks, that love of liberty which had been their most glorious distinction among the nations of mankind; and Alexander became the instrument of its completion. The mischiefs attendant on this consummation of fruitless ruin are too manifold and too tremendous to be related. If all the thought which had been expended on the construction of engines of agony and death—the modes of aggression and defence, the raising of armies, and the acquirement of those arts of tyranny and falsehood without which mixed multitudes could neither be led nor governed—had been employed to promote the true welfare and extend the real empire of man, how different would have been the present situation of human society! how different the state of knowledge in physical and moral science, upon which the power and happiness of mankind essentially depend ! What nation has the example of the desolation of Attica by Mardonius and Xerxes, or the extinction of the Persian empire by Alexander of Macedon, restrained from outrage ? Was not the pretext of this latter system of spoliation derived immediately from the former ? Had revenge in this instance any other effect than to increase, instead of diminishing, the mass of malice and evil already existing in the world ?

The emptiness and folly of retaliation are apparent from every example which can be brought forward. Not only Jesus Christ, but the most eminent professors of every sect of philosophy, have reasoned against this futile superstition. Legislation is, in one point of view, to be considered as an attempt to provide against the excesses of this deplorable mistake. It professes to assign the penalty of all private injuries, and denies to individuals the right of vindicating their proper cause. This end is certainly not attained without some accommodation to the propensities which it desires to destroy. Still, it recognizes no principle but the production of the greatest eventual good with the least immediate injury ; and regards the torture, or the death, of any human being as unjust, of whatever mischief he may have been the author, so that the result shall not more than compensate for the immediate pain.

Mankind, transmitting from generation to generation the legacy of accumulated vengeances, and pursuing with the feelings of duty the misery of their fellow beings, have not failed to attribute to the Universal Cause a character analogous with their own. The image of this invisible, mysterious Being is more or less excellent and perfect—resembles more or less its original—in proportion to the perfection of the mind on which it is impressed. Thus, that nation which has arrived at the highest step in the scale of moral progression will believe most purely in that God, the knowledge of whose real attributes is considered as the firmest basis of the true religion. The reason of the belief of each individual, also, will be so far regulated by his conceptions of what is good. Thus, the conceptions which any nation or individual entertains of the God of its popular worship may be inferred from their own actions and opinions, which are the subjects of their approbation among their fellow men. Jesus Christ instructed his disciples to be perfect, as their Father in Heaven is perfect, declaring at the same time his belief that human perfection requires the refraining from revenge and retribution in any of its various shapes.

The perfection of the human and the divine character is thus asserted to be the same. Man, by resembling God, fulfils most accurately the tendencies of his nature ; and God comprehends within himself all that constitutes human perfection. Thus, God is a model through which the excellence of man is to be estimated, whilst the *abstract* perfection of the human character is the type of the *actual* perfection of the divine. It is not to be believed that a person of such comprehensive views as Jesus Christ could have fallen into so manifest a contradiction as to assert that men would be tortured after death by that Being whose character is held up as a model to human kind, because he is incapable of malevolence and revenge. All the arguments which have been brought forward to justify retribution fail, when retribution is destined neither to operate as an example to other agents, nor to the offender himself. How feeble such reasoning is to be considered, has been already shown ; but it is the character of an evil Daemon to consign the beings whom he has endowed with sensation to unprofitable anguish. The peculiar circumstances attendant on the conception of God casting sinners to burn in Hell for ever, combine to render that conception the most perfect specimen of the greatest imaginable crime. Jesus Christ represented God as the principle of all good, the source of all happiness, the wise and benevolent Creator and Preserver of all living things. But the interpreters of his doctrines have confounded the good and the evil principle. They observed the emanations of their universal natures to be inextricably entangled in the world, and, trembling before the power of the cause of all things, addressed to it such flattery as is acceptable to the ministers of human tyranny, attributing love and wisdom to those energies which they felt to be exerted indifferently for the purposes of benefit and calamity.

Jesus Christ expressly asserts that distinction between the good and evil principle which it has been the practice of all theologians to confound. How far his doctrines, or their interpretation, may be true, it would scarcely have been worth while to inquire, if the one did not afford an example and an incentive to the attainment of true virtue, whilst the other holds out a sanction and apology for every species of mean and cruel vice.

It cannot be precisely ascertained in what degree Jesus Christ accommodated his doctrines to the opinions of his auditors; or in what degree he really said all that he is related to have said. He has left no written record of himself, and we are compelled to judge from the imperfect and obscure information which his biographers (persons certainly of very undisciplined and undiscriminating minds) have transmitted to posterity. These writers (our only guides) impute sentiments to Jesus Christ which flatly contradict each other. They represent him as narrow, superstitious, and exquisitely vindictive and malicious. They insert, in the midst of a strain of impassioned eloquence or sagest exhortation, a sentiment only remarkable for its naked and drivelling folly. But it is not difficult to distinguish the inventions by which these historians have filled up the interstices of tradition, or corrupted the simplicity of truth, from the real character of their rude amazement. They have left sufficiently clear indications of the genuine character of Jesus Christ to rescue it for ever from the imputations cast upon it by their ignorance and fanaticism. We discover that he is the enemy of oppression and of falsehood ; that he is the advocate of equal justice ; that he is neither disposed to sanction bloodshed nor deceit, under whatsoever pretences their practice may be vindicated. We discover that he was a man of meek and majestic demeanour, calm in danger ; of natural and simple thought and habits ; beloved to adoration by his adherents ; unmoved, solemn, and severe.

It is utterly incredible that this man said, that if you hate your enemy, you would find it to your account to return him good for evil, since, by such a temporary oblivion of vengeance, you would heap coals of fire on his head. Where such contradictions occur, a favourable construction is warranted by the general innocence of manners and comprehensiveness of views which he is represented to possess. The rule of criticism to be adopted in judging of the life, actions, and words of a man who has acted any conspicuous part in the revolutions of the world, should not be narrow. We ought to form a general image of his character and of his doctrines, and refer to this whole the distinct portions of actions and speech by which they are diversified. It is not here asserted that no contradictions are to be admitted to have taken place in the system of Jesus Christ, between doctrines promulgated in different states of feeling or information, or even such as are implied in the enunciation of a scheme of thought, various and obscure through its immensity and depth. It is not asserted that no degree of human indignation ever hurried him, beyond the limits which his calmer mood had placed, to disapprobation against vice and folly. Those deviations from the history of his life are alone to be vindicated, which represent his own essential character in contradiction with itself.

Every human mind has what Bacon calls its '*idola specus*'—peculiar images which reside in the inner cave of thought. These constitute the essential and distinctive character of every human being; to which every action and every word have intimate relation; and by which, in depicting a character, the genuineness and meaning of these words and actions are to be determined. Every fanatic or enemy of virtue is not at liberty to misrepresent the greatest geniuses and most heroic defenders of all that is valuable in this mortal world. History, to gain any credit, must contain some truth, and that truth shall thus be made a sufficient indication of prejudice and deceit.

With respect to the miracles which these biographers have related, I have already declined to enter into any discussion on their nature or their existence. The supposition of their falsehood or their truth would modify in no degree the hues of the picture which is attempted to be delineated. To judge truly of the moral and philosophical character of Socrates, it is not necessary to determine the question of the familiar Spirit which [it] is supposed that he believed to attend on him. The power of the human mind, relatively to intercourse with or dominion over the invisible world, is doubtless an interesting theme of discussion; but the connexion of the instance of Jesus Christ with the established religion of the country in which I write, renders it dangerous to subject oneself to the imputation of introducing new Gods or abolishing old ones; nor is the duty of mutual forbearance sufficiently understood to render it certain that the metaphysician and the moralist, even though he carefully sacrifice a cock to Esculapius, may not receive something analogous to the bowl of hemlock for the reward of his labours. Much, however, of what his [Christ's] biographers have asserted is not to be rejected merely because inferences inconsistent with the general spirit of his system are to be adduced from its admission. Jesus Christ did what every other reformer who has produced any considerable effect upon the world has done. He accommodated his doctrines to the prepossessions of those whom he addressed. He used a language for this view sufficiently familiar to our comprehensions. He said,—However new or strange my doctrines may appear to you, they are in fact only the restoration and re-establishment of those original institutions and ancient customs of your own law and religion. The constitutions of your faith and policy, although perfect in their origin, have become corrupt and altered, and have fallen into decay. I profess to restore them to their pristine authority and splendour. 'Think not that I am come to destroy the Law and the Prophets. I am come not to destroy, but to fulfil. Till heaven and earth pass away, one jot or one tittle shall in nowise pass away from the Law, till all be fulfilled.' Thus, like a skilful orator (see Cicero, *De Oratore*), he secures the prejudices of his auditors, and induces them, by his professions of sympathy with their feelings, to enter with a willing mind into the exposition of his own. The art of persuasion differs from that of reasoning; and it is of no small moment, to the success even of a true cause, that the judges who are to determine on its merits should be free from those national and religious predilections which render the multitude both deaf and blind.

Let not this practice be considered as an unworthy artifice. It were best for the cause of reason that mankind should acknowledge no authority but its own; but it is useful, to a certain extent, that then should not consider those institutions which they have been habituated to reverence as opposing ay obstacle to its admission. All reformers have been compelled to practise this misrepresentation of their own true feelings and opinions. It is deeply to be lamented that a word should ever issue from human lips which contains the minutest alloy of dissimulation, or simulation, or hypocrisy, or exaggeration, or anything but the precise and rigid image which is present to the mind, and which ought to dictate the expression. But the practice of utter sincerity towards other men would avail to no good end, if they were incapable of practising it towards their own minds. In fact, truth cannot be communicated until it is perceived. The interests, therefore, of truth require that an orator should, as far as possible, produce in his hearers that state of mind on which alone his exhortations could fairly be contemplated and examined.

Having produced this favourable disposition of mind, Jesus Christ proceeds to qualify, and finally to abrogate, the system of the Jewish law. He descants upon its insufficiency as a code of moral conduct, which it professed to be, and absolutely selects the law of retaliation as an instance of the absurdity and immorality of its institutions. The conclusion of the speech is in a strain of the most daring and most impassioned speculation. He seems emboldened by the success of his exculpation to the multitude, to declare in public the utmost singularity of his faith. He tramples upon all received opinions, on all the cherished luxuries and superstitions of mankind. He bids them cast aside the claims of custom and blind faith by which they have been encompassed from the very cradle of their being, and receive the imitator and minister of the Universal God.

### EQUALITY OF MANKIND

'The spirit of the Lord is upon me, because he hath chosen me to preach the gospel to the poor: He hath sent me to heal the broken-hearted, to preach deliverance to the captives and recovery of sight to the blind, and to set at liberty them that are bruised' (Luke iv. 18). This is an enunciation of all that Plato and Diogenes have speculated upon the equality of mankind. They saw that the great majority of the human species were reduced to the situation of squalid ignorance and moral imbecility, for the purpose of purveying for the luxury of a few, and contributing to the satisfaction of their thirst for power. Too mean-spirited and too feeble in resolve to attempt the conquest of their own evil passions and of the difficulties of the material world, men sought dominion over their fellow men, as an easy method to gain that apparent majesty and power which the instinct of their nature requires. Plato wrote the scheme of a republic, in which law should watch over the equal distribution of the external instruments of unequal power—honours, property, &c. Diogenes devised a nobler and a more worthy system of opposition to the system of the slave and tyrant. He said: 'It is in the power of each individual to level the inequality which is the topic of the complaint of mankind. Let him be aware of his own worth, and the station which he occupies in the scale of moral beings. Diamonds and gold, palaces and sceptres, derive their value from the opinion of mankind. The only sumptuary law which can be imposed on the use and fabrication of these instruments of mischief and deceit, these symbols of successful injustice, is the law of opinion. Every man possesses the power, in this respect, to legislate for himself. Let him be well aware of his own worth and moral dignity. Let him yield in meek reverence to any wiser or worthier than he, so long as he accords no veneration to the splendour of his apparel, the luxury of his food, the multitude of his flatterers and slaves. It is because, mankind, ye value and seek the empty pageantry of wealth and social power, that ye are enslaved to its possessions. Decrease your physical wants; learn to live, so far as nourishment and shelter are concerned, like the beast of the forest and the birds of the air; ye will need not to complain, that other individuals of your species are surrounded by the diseases of luxury and the vices of subserviency and oppression.' With all those who are truly wise, there will be an entire community, not only of thoughts and feelings, but also of external possessions. Insomuch, therefore, as ye live [wisely], ye may enjoy the community of whatsoever benefits arise from the inventions of civilized life. They are of value only for purposes of mental power; they are of value only as they are capable of being shared and applied to the common advantage of philosophy; and if there be no love among men, whatever institutions they may frame must be subservient to the same purpose—to the continuance of inequality. If there be no love among men, it is best that he who sees through the hollowness of their professions should fly from their society, and suffice to his own soul. In wisdom, he will thus lose nothing ; in power, he will gain everything. In proportion to the love existing among men, so will be the community of property and power. Among true and real friends, all is common ; and, were ignorance and envy and superstition banished from the world, all mankind would be friends. The only perfect and genuine republic is that which comprehends every living being. Those distinctions which have been artificially set up, of nations, societies, families, and religions, are only general names, expressing the abhorrence and contempt with which men blindly consider their fellow men. I love my country ; I love the city in which I was born, my parents, my wife, and the children of my care ; and to this city, this woman, and this nation, it is incumbent on me to do all the benefit in my power. To what do these distinctions point, but to an evident denial of the duty which humanity imposes on you, of doing every possible good to every individual, under whatever denomination he may be comprehended, to whom you have the power of doing it ? You ought to love all mankind ; nay, every individual of mankind. You ought not to love the individuals of your domestic circle less, but to love those who exist beyond it more. Once make the feelings of confidence and of affection universal, and the distinctions of property and power will vanish ; nor are they to be abolished without substituting something equivalent in mischief to them, until all mankind shall acknowledge an entire community of rights.

But, as the shades of night are dispelled by the

faintest glimmerings of dawn, so shall the minutest progress of the benevolent feelings disperse, in some degree, the gloom of tyranny, and [curb the] ministers of mutual suspicion and abhorrence. Your physical wants are few, whilst those of your mind and heart cannot be numbered or described, from their multitude and complication. To secure the gratification of the former, you have made yourselves the bond-slaves of each other.

They have cultivated these meaner wants to so great an excess as to judge nothing so valuable or desirable [as] what relates to their gratification. Hence has arisen a system of passions which loses sight of the end they were originally awakened to attain. Fame, power, and gold, are loved for their own sakes—are worshipped with a blind, habitual idolatry. The pageantry of empire, and the fame of irresistible might, are contemplated by the possessor with unmeaning complacency, without a retrospect to the properties which first made him consider them of value. It is from the cultivation of the most contemptible properties of human nature that discord and torpor and indifference, by which the moral universe is disordered, essentially depend. So long as these are the ties by which human society is connected, let it not be admitted that they are fragile.

Before man can be free, and equal, and truly wise, he must cast aside the chains of habit and superstition; he must strip sensuality of its pomp, and selfishness of its excuses, and contemplate actions and objects as they really are. He will discover the wisdom of universal love; he will feel the meanness and the injustice of sacrificing the reason and the liberty of his fellow men to the indulgence of his physical appetites, and becoming a party to their degradation by the consummation of his own.

Such, with those differences only incidental to the age and state of society in which they were promulgated, appear to have been the doctrines of Jesus Christ. It is not too much to assert that they have been the doctrines of every just and compassionate mind that ever speculated on the social nature of man. The dogma of the equality of mankind has been advocated, with various success, in different ages of the world. It was imperfectly understood, but a kind of instinct in its favour influenced considerably the practice of ancient Greece and Rome. Attempts to establish usages founded on this dogma have been made in modern Europe, in several instances, since the revival of literature and the arts. Rousseau has vindicated this opinion with all the eloquence of sincere and earnest faith ; and is, perhaps, the philosopher among the moderns who, in the structure of his feelings and understanding, resembles most nearly the mysterious sage of Judea. It is impossible to read those passionate words in which Jesus Christ upbraids the pusillanimity and sensuality of mankind, without being strongly reminded of the more connected and systematic enthusiasm of Rousseau. 'No man,' says Jesus Christ, 'can serve two masters. Take, therefore, no thought for to-morrow, for the morrow shall take thought for the things of itself. Sufficient unto the day is the evil thereof.' If we would profit by the wisdom of a sublime and poetical mind, we must beware of the vulgar error of interpreting literally every expression it employs. Nothing can well be more remote from truth than the literal and strict construction of such expressions as Jesus Christ delivers, or than [to imagine that] it were best for man that he should abandon all his acquirements in physical and intellectual science, and depend on the spontaneous productions of nature for his subsistence. Nothing is more obviously false than that the remedy for the inequality among men consists in their return to the condition of savages and beasts. Philosophy will never be understood if we approach the study of its mysteries with so narrow and illiberal conceptions of its universality. Rousseau certainly did not mean to persuade the immense population of his country to abandon all the arts of life, destroy their habitations and their temples, and become the inhabitants of the woods. He addressed the most enlightened of his compatriots, and endeavoured to persuade them to set the example of a pure and simple life, by placing in the strongest point of view his conceptions of the calamitous and diseased aspect which, overgrown as it is with the vices of sensuality and selfishness, is exhibited by civilized society. Nor can it be believed that Jesus Christ endeavoured to prevail on the inhabitants of Jerusalem neither to till their fields, nor to frame a shelter against the sky, nor to provide food for the morrow. He simply exposes, with the passionate rhetoric of enthusiastic love towards all human beings, the miseries and mischiefs of that system which makes all things subservient to the subsistence of the material frame of man. He warns them that no man can serve two masters—God and Mammon ; that it is impossible at once to be high-minded and just and wise, and to comply with the accustomed forms of human society, seek power, wealth, or empire, either from the idolatry of habit, or as the direct instruments of sensual gratification. He instructs them that clothing and food and shelter are not, as they suppose, the true end of human life, but only certain means, to be valued in proportion to their subserviency to that end. These means it is right of every human being to possess, and that in the same degree. In this respect, the fowls of the air and the lilies of the field are examples for the imitation of mankind. They are clothed and fed by the Universal God. Permit, therefore, the Spirit of this benignant Principle to visit your intellectual frame, or, in other words, become just and pure. When you understand the degree of attention which the requisitions of your physical nature demand, you will perceive how little labour suffices for their satisfaction. Your Heavenly Father knoweth you have need of these things. The universal Harmony, or Reason, which makes your passive frame of thought its dwelling, in proportion to the purity and majesty of its nature will instruct you, if ye are willing to attain that exalted condition, in what manner to possess all the objects necessary for your material subsistence. All men are [impelled] to become thus pure and happy. All men are called to participate in the community of Nature's gifts. The man who has fewest bodily wants approaches nearest to the Divine Nature. Satisfy these wants at the cheapest rate, and expend the remaining energies of your nature in the attainment of virtue and knowledge. The mighty frame of the wonderful and lovely world is the food of your contemplation, and living beings who resemble your own nature, and are bound to you by similarity of sensations, are destined to be the nutriment of your affection; united, they are the consummation of the widest hopes your mind can contain. Ye can expend thus no labour on mechanism consecrated to luxury and pride. How abundant will not be your progress in all that truly ennobles and extends human nature! By rendering yourselves thus worthy, ye will be as free in your imaginations as the swift and many-coloured fowls of the air, and as beautiful in pure simplicity as the lilies of the field. In proportion as mankind becomes wise—yes, in exact proportion to that wisdom—should be the extinction of the unequal system under which they now subsist. Government is, in fact, the mere badge of their depravity. They are so little aware of the inestimable benefits of mutual love as to indulge, without thought, and almost without motive, in the worst excesses of selfishness and malice. Hence, without graduating human society into a scale of empire and subjection, its very existence has become impossible. It is necessary that universal benevolence should supersede the regulations of precedent and prescription, before these regulations can safely be abolished. Meanwhile, their very subsistence depends on the system of injustice and violence which they have been devised to palliate. They suppose men endowed with the power of deliberating and determining for their equals; whilst these men, as frail and as ignorant as the multitude whom they rule, possess, as a practical consequence of this power, the right which they of necessity exercise to prevent (together with their own) the physical and moral and intellectual nature of all mankind.

It is the object of wisdom to equalize the distinctions on which this power depends, by exhibiting in their proper worthlessness the objects, a contention concerning which renders its existence a necessary evil. The evil, in fact, is virtually abolished wherever *justice* is practised; and it is abolished in precise proportion to the prevalence of true virtue.

The whole frame of human things is infected by an insidious poison. Hence it is that man is blind in his understanding, corrupt in his moral sense, and diseased in his physical functions. The wisest and most sublime of the ancient poets saw this truth, and embodied their conception of its value in retrospect to the earliest ages of mankind. They represented equality as the reign of Saturn, and taught that mankind had gradually degenerated from the virtue which enabled them to enjoy or maintain this happy state. Their doctrine was philosophically false. Later and more correct observations have instructed us that uncivilized man is the most pernicious and miserable of beings, and that the violence and injustice, which are the genuine indications of real inequality, obtain in the society of these beings without palliation. Their imaginations of a happier state of human society were referred, in truth, to the Saturnian period; they ministered, indeed, to thoughts of despondency and sorrow. But they were the children of airy hope—the prophets and parents of man's futurity. Man was once as a wild beast; he has become a moralist, a metaphysician, a poet, and an astronomer. Lucretius or Virgil might have referred the comparison to themselves; and, as a proof of the progress of the nature of man, challenged a comparison with the cannibals of Scythia.[1] The experience of the ages which have intervened between the present period and that in which Jesus Christ taught, tends to prove his doctrine, and to illustrate theirs. There is more equality because there is more justice, and there is more justice because there is more universal knowledge.

To the accomplishment of such mighty hopes were the views of Jesus Christ extended; such did he believe to be the tendency of his doctrines—the abolition of artificial distinctions among mankind, so far as the love which it becomes all human beings to bear towards each other, and the knowledge of truth from which that love will never fail to be produced, avail to their destruction. A young man came

[1] Jesus Christ foresaw what the poets retrospectively imagined.

to Jesus Christ, struck by the miraculous dignity and simplicity of his character, and attracted by the words of power which he uttered. He demanded to be considered as one of the followers of his creed. 'Sell all that thou hast,' replied the philosopher; 'give it to the poor, and follow me.' But the young man had large possessions, and he went away sorrowing.

The system of equality was attempted, after Jesus Christ's death, to be carried into effect by his followers. 'They that believed had all things in common; they sold their possessions and goods, and parted them to all men, as every man had need; and they continued daily with one accord in the temple, and, breaking bread from house to house, did eat their meat with gladness and singleness of heart.' (Acts ii.)

The practical application of the doctrines of strict justice to a state of society established in its contempt, was such as might have been expected. After the transitory glow of enthusiasm had faded from the minds of men, precedent and habit resumed their empire; they broke like a universal deluge on one shrinking and solitary island. Men to whom birth had allotted ample possession, looked with complacency on sumptuous apartments and luxurious food, and those ceremonials of delusive majesty which surround the throne of power and the court of wealth. Men from whom these things were withheld by their condition, began again to gaze with stupid envy on pernicious splendour; and, by desiring the false greatness of another's state, to sacrifice the intrinsic dignity of their own. The demagogues of the infant republic of the Christian sect, attaining, through eloquence or artifice, to influence amongst its members, first violated (under the pretence of watching over their integrity) the institutions established for the common and equal benefit of all. These demagogues artfully silenced the voice of the moral sense among them by engaging them to attend, not so much to the cultivation of a virtuous and happy life in this mortal scene, as to the attainment of a fortunate condition after death ; not so much to the consideration of those means by which the state of man is adorned and improved, as an inquiry into the secrets of the connexion between God and the world—things which, they well knew, were not to be explained, or even to be conceived. The system of equality which they established necessarily fell to the ground, because it is a system that must result from, rather than precede, the moral improvement of human kind. It was a circumstance of no moment that the first adherents of the system of Jesus Christ cast their property into a common stock. The same degree of real community of property could have subsisted without this formality, which served only to extend a temptation of dishonesty to the treasurers of so considerable a patrimony. Every man, in proportion to his virtue, considers himself, with respect to the great community of mankind, as the steward and guardian of their interests in the property which he chances to possess. Every man, in proportion to his wisdom, sees the manner in which it is his duty to employ the resources which the consent of mankind has entrusted to his discretion. Such is the [annihilation] of the unjust inequality of powers and conditions existing in the world ; and so gradually and inevitably is the progress of equality accommodated to the progress of wisdom and of virtue among mankind.

Meanwhile, some benefit has not failed to flow from the imperfect attempts which have been made to erect a system of equal rights to property and power upon the basis of arbitrary institutions. They have undoubtedly, in every case, from the instability of their formation, failed. Still, they constitute a record of those epochs at which a true sense of justice suggested itself to the understandings of men, so that they consented to forgo all the cherished delights of luxury, all the habitual gratifications arising out of the possession or the expectation of power, all the superstitions with which the accumulated authority of ages had made them dear and venerable. They are so many trophies erected in the enemy's land, to mark the limits of the victorious progress of truth and justice.

Jesus Christ did not fail to advert to the——

[THE REST IS WANTING.]

1815

# ON THE REVIVAL OF LITERATURE

In the fifteenth century of the Christian era, a new and extraordinary event roused Europe from her lethargic state, and paved the way to her present greatness. The writings of Dante in the thirteenth, and of Petrarch in the fourteenth, were the bright luminaries which had afforded glimmerings of literary knowledge to the almost benighted traveller toiling up the hill of Fame. But on the taking of Constantinople, a new and sudden light appeared: the dark clouds of ignorance rolled into distance, and Europe was inundated by learned monks, and still more by the quantity of learned manuscripts which they brought with them from the scene of devastation. The Turks settled themselves in Constantinople, where they adopted nothing but the vicious habits of the Greeks: they neglected even the small remains of its ancient learning, which, filtered and degenerated as it was by the absurd mixture of Pagan and Christian philosophy, proved, on its retirement to Europe, the spark which spread gradually and successfully the light of knowledge over the world.

Italy, France, and England—for Germany still remained many centuries less civilized than the surrounding countries,—swarmed with monks and cloisters. Superstition, of whatever kind, whether earthly or divine, has hitherto been the weight which clogged man to earth, and prevented his genius from soaring aloft amid its native skies. The enterprises, and the effects of the human mind, are something more than stupendous: the works of nature are material and tangible: we have a half insight into their kind, and in many instances we predict their effects with certainty. But mind seems to govern the world without visible or substantial means. Its birth is unknown; its action and influence unperceived; and its being seems eternal. To the mind both humane and philosophical, there cannot exist a greater subject of grief, than the reflection of how much superstition has retarded the progress of intellect, and consequently the happiness of man.

The monks in their cloisters were engaged in trifling and ridiculous disputes: they contented themselves with teaching the dogmas of their religion, and rushed impatiently forth to the colleges and halls, where they disputed with an acrimony and meanness little befitting the resemblance of their pretended holiness. But the situation of a monk is a situation the most unnatural that bigotry, proud in the invention of cruelty, could conceive; and their vices may be pardoned as resulting from the wills[1] and devices of a few proud and selfish bishops, who enslaved the world that they might live at ease.

[1] ? wiles [Ed.].

The disputes of the schools were mostly scholastical; it was the discussion of words, and had no relation to morality. Morality,—the great means and end of man,—was contained, as they affirmed, in the extent of a few hundred pages of a certain book, which others have since contended were but scraps of martyrs' last dying words, collected together and imposed on the world. In the refinements of the scholastic philosophy, the world seemed in danger of losing the little real wisdom that still remained as her portion; and the only valuable part of their disputes was such as tended to develop the system of the Peripatetic Philosophers. Plato, the wisest, the profoundest, and Epicurus, the most humane and gentle among the ancients, were entirely neglected by them. Plato interfered with their peculiar mode of thinking concerning heavenly matters; and Epicurus, maintaining the rights of man to pleasure and happiness, would have afforded a seducing contrast to their dark and miserable code of morals. It has been asserted, that these holy men solaced their lighter moments in a contraband worship of Epicurus and profaned the philosophy which maintained the rights of all by a selfish indulgence of the rights of a few. Thus it is: the laws of nature are invariable, and man sets them aside that he may have the pleasure of travelling through a labyrinth in search of them again.

Pleasure, in an open and innocent garb, by some strange process of reasoning, is called vice; yet man (so closely is he linked to the chains of necessity—so irresistibly is he impelled to fulfil the end of his being,) must seek her at whatever price: he becomes a hypocrite, and braves damnation with all its pains.

Grecian literature,—the finest the world has ever produced,—was at length restored: its form and mode we obtained from the manuscripts which the ravages of time, of the Goths, and of the still more savage Turks, had spared. The burning of the library at Alexandria was an evil of importance. This library is said to have contained volumes of the choicest Greek authors.

1815

# A DEFENCE OF POETRY

## PART I

According to one mode of regarding those two classes of mental action, which are called reason and imagination, the former may be considered as mind contemplating the relations borne by one thought to another, however produced; and the latter, as mind acting upon those thoughts so as to colour them with its own light, and composing from them, as from elements, other thoughts, each containing within itself the principle of its own integrity. The one is the *τὸ ποιεῖν*, or the principle of synthesis, and has for its objects those forms which are common to universal nature and existence itself; the other is the *τὸ λογίζειν*, or principle of analysis, and its action regards the relations of things, simply as relations; considering thoughts, not in their integral unity, but as the algebraical representations which conduct to certain general results. Reason is the enumeration of quantities already known; imagination is the perception of the value of those quantities, both separately and as a whole. Reason respects the differences, and imagination the similitudes of things. Reason is to the imagination as the instrument to the agent, as the body to the spirit, as the shadow to the substance.

Poetry, in a general sense, may be defined to be 'the expression of the imagination': and poetry is connate with the origin of man. Man is an instrument over which a series of external and internal impressions are driven, like the alternations of an ever-changing wind over an Aeolian lyre, which move it by their motion to ever-changing melody. But there is a principle within the human being, and perhaps within all sentient beings, which acts otherwise than in the lyre, and produces not melody alone, but harmony, by an internal adjustment of the sounds or motions thus excited to the impressions which excite them. It is as if the lyre could accommodate its chords to the motions of that which strikes them, in a determined proportion of sound; even as the musician can accommodate his voice to the sound of the lyre. A child at play by itself will express its delight by its voice and motions; and every inflexion of tone and every gesture will bear exact relation to a corresponding antitype in the pleasurable impressions which awakened it; it will be the reflected image of that impression; and as the lyre trembles and sounds after the wind has died away, so the child seeks, by prolonging in its voice and motions the duration of the effect, to prolong also a consciousness of the cause. In relation to the objects which delight a child, these expressions are, what poetry is to higher objects. The savage (for the savage is to ages what the child is to years) expresses the emotions produced in him by surrounding objects in a similar manner; and language and gesture, together with plastic or pictorial imitation, become the image of the combined effect of those objects, and of his apprehension of them. Man in society, with all his passions and his pleasures, next becomes the object of the passions and pleasures of man; an additional class of emotions produces an augmented treasure of expressions ; and language, gesture, and the imitative arts, become at once the representation and the medium, the pencil and the picture, the chisel and the statue, the chord and the harmony. The social sympathies, or those laws from which, as from its elements, society results, begin to develop themselves from the moment that two human beings coexist ; the future is contained within the present, as the plant within the seed ; and equality, diversity, unity, contrast, mutual dependence, become the principles alone capable of affording the motives according to which the will of a social being is determined to action, inasmuch as he is social ; and constitute pleasure in sensation, virtue in sentiment, beauty in art, truth in reasoning, and love in the intercourse of kind. Hence men, even in the infancy of society, observe a certain order in their words and actions, distinct from that of the objects and the impressions represented by them, all expression being subject to the laws of that from which it proceeds. But let us dismiss those more general considerations which might involve an inquiry into the principles of society itself, and restrict our view to the manner in which the imagination is expressed upon its forms.

In the youth of the world, men dance and sing and imitate natural objects, observing in these actions, as in all others, a certain rhythm or order. And, although all men observe a similar, they observe not the same order, in the motions of the dance, in the melody of the song, in the combinations of language, in the series of their imitations of natural objects. For there is a certain order or rhythm belonging to each of these classes of mimetic representation, from which the hearer and the spectator receive an intenser and purer pleasure than from any other : the sense of an approximation to this order has been called taste by modern writers. Every man in the infancy of art observes an order which approximates more or less closely to that from which this highest delight results : but the diversity is not sufficiently marked, as that its gradations should be sensible, except in those instances where the predominance of this faculty of approximation to the beautiful (for so we may be permitted to name the relation between this highest pleasure and its cause) is very great. Those in whom it exists in excess are poets, in the most universal sense of the word ; and the pleasure resulting from the manner in which they express the influence of society or nature upon their own minds, communicates itself to others, and gathers a sort of reduplication from that community. Their language is vitally metaphorical ; that is, it marks the before unapprehended relations of things and perpetuates their apprehension, until the words which represent them become, through time, signs for portions or classes of thoughts instead of pictures of integral thoughts ; and then if no new poets should arise to create afresh the associations which have been thus disorganized, language will be dead to all the nobler purposes of human intercourse. These similitudes or relations are finely said by Lord Bacon to be 'the same footsteps of nature impressed upon the various subjects of the world' ;[1] and he considers the faculty which perceives them as the storehouse of axioms common to all knowledge. In the infancy of society every author is necessarily a poet, because language itself is poetry ; and to be a poet is to apprehend the true and the beautiful, in a word, the good which exists in the relation, subsisting, first between existence and perception, and secondly between perception and expression. Every

[1] *De Augment. Scient.*, cap. i, lib. iii.

original language near to its source is in itself the chaos of a cyclic poem: the copiousness of lexicography and the distinctions of grammar are the works of a later age, and are merely the catalogue and the form of the creations of poetry.

But poets, or those who imagine and express this indestructible order, are not only the authors of language and of music, of the dance, and architecture, and statuary, and painting; they are the institutors of laws, and the founders of civil society, and the inventors of the arts of life, and the teachers, who draw into a certain propinquity with the beautiful and the true, that partial apprehension of the agencies of the invisible world which is called religion. Hence all original religions are allegorical, or susceptible of allegory, and, like Janus, have a double face of false and true. Poets, according to the circumstances of the age and nation in which they appeared, were called, in the earlier epochs of the world, legislators, or prophets: a poet essentially comprises and unites both these characters. For he not only beholds intensely the present as it is, and discovers those laws according to which present things ought to be ordered, but he beholds the future in the present, and his thoughts are the germs of the flower and the fruit of latest time. Not that I assert poets to be prophets in the gross sense of the word, or that they can foretell the form as surely as they foreknow the spirit of events: such is the pretence of superstition, which would make poetry an attribute of prophecy, rather than prophecy an attribute of poetry. A poet participates in the eternal, the infinite, and the one; as far as relates to his conceptions, time and place and number are not. The grammatical forms which express the moods of time, and the difference of persons, and the distinction of place, are convertible with respect to the highest poetry without injuring it as poetry ; and the choruses of Aeschylus, and the book of *Job*, and Dante's *Paradise,* would afford, more than any other writings, examples of this fact, if the limits of this essay did not forbid citation. The creations of sculpture, painting, and music, are illustrations still more decisive.

Language, colour, form, and religious and civil habits of action, are all the instruments and materials of poetry ; they may be called poetry by that figure of speech which considers the effect as a synonym of the cause. But poetry in a more restricted sense expresses those arrangements of language, and especially metrical language, which are created by that imperial faculty, whose throne is curtained within the invisible nature of man. And this springs from the nature itself of language, which is a more direct representation of the actions and passions of our internal being, and is susceptible of more various and delicate combinations, than colour, form, or motion, and is more plastic and obedient to the control of that faculty of which it is the creation. For language is arbitrarily produced by the imagination, and has relation to thoughts alone ; but all other materials, instruments, and conditions of art, have relations among each other, which limit and interpose between conception and expression. The former is as a mirror which reflects, the latter as a cloud which enfeebles, the light of which both are mediums of communication. Hence the fame of sculptors, painters, and musicians, although the intrinsic powers of the great masters of these arts may yield in no degree to that of those who have employed language as the hieroglyphic of their thoughts, has never equalled that of poets in the restricted sense of the term ; as two performers of equal skill will produce unequal effects from a guitar and a harp. The fame of legislators and founders of religions, so long as their institutions last, alone seems to exceed that of poets in the restricted sense ; but it can scarcely be a question, whether, if we deduct the celebrity which their flattery of the gross opinions of the vulgar usually conciliates, together with that which belonged to them in their higher character of poets, any excess will remain.

We have thus circumscribed the word poetry within the limits of that art which is the most familiar and the most perfect expression of the faculty itself. It is necessary, however, to make the circle still narrower, and to determine the distinction between measured and unmeasured language ; for the popular division into prose and verse is inadmissible in accurate philosophy.

Sounds as well as thoughts have relation both between each other and towards that which they represent, and a perception of the order of those relations has always been found connected with a perception of the order of the relations of thoughts. Hence the language of poets has ever affected a certain uniform and harmonious recurrence of sound, without which it were not poetry, and which is scarcely less indispensable to the communication of its influence, than the words themselves, without reference to that peculiar order. Hence the vanity of translation ; it were as wise to cast a violet into a crucible that you might discover the formal principle of its colour and odour, as seek to transfuse from one language into another the creations of a poet. The plant must spring again from its seed, or it will bear no flower—and this is the burthen of the curse of Babel.

An observation of the regular mode of the recurrence of harmony in the language of poetical minds, together with its relation to music, produced metre, or a certain system of traditional forms of harmony and language. Yet it is by no means essential that a poet should accommodate his language to this traditional form, so that the harmony, which is its spirit, be observed. The practice is indeed convenient and popular, and to be preferred, especially in such composition as includes much action: but every great poet must inevitably innovate upon the example of his predecessors in the exact structure of his peculiar versification. The distinction between poets and prose writers is a vulgar error. The distinction between philosophers and poets has been anticipated. Plato was essentially a poet—the truth and splendour of his imagery, and the melody of his language, are the most intense that it is possible to conceive. He rejected the measure of the epic, dramatic, and lyrical forms, because he sought to kindle a harmony in thoughts divested of shape and action, and he forbore to invent any regular plan of rhythm which would include, under determinate forms, the varied pauses of his style. Cicero sought to imitate the cadence of his periods, but with little success. Lord Bacon was a poet.[1] His language has a sweet and majestic rhythm, which satisfies the sense, no less than the almost superhuman wisdom of his philosophy satisfies the intellect; it is a strain which distends, and then bursts the circumference of the reader's mind, and pours itself forth together with it into the universal element with which it has perpetual sympathy. All the authors of revolutions in opinion are not only necessarily poets as they are inventors, nor even as their words unveil the permanent analogy of things by images which participate in the life of truth; but as their periods are harmonious and rhythmical, and contain in themselves the elements of verse; being the echo of the eternal music. Nor are those supreme poets, who have employed traditional forms of rhythm on account

[1] See the *Filum Labyrinthi*, and the Essay on Death particularly.

of the form and action of their subjects, less capable of perceiving and teaching the truth of things, than those who have omitted that form. Shakespeare, Dante, and Milton (to confine ourselves to modern writers) are philosophers of the very loftiest power.

A poem is the very image of life expressed in its eternal truth. There is this difference between a story and a poem, that a story is a catalogue of detached facts, which have no other connexion than time, place, circumstance, cause and effect ; the other is the creation of actions according to the unchangeable forms of human nature, as existing in the mind of the Creator, which is itself the image of all other minds. The one is partial, and applies only to a definite period of time, and a certain combination of events which can never again recur ; the other is universal, and contains within itself the germ of a relation to whatever motives or actions have place in the possible varieties of human nature. Time, which destroys the beauty and the use of the story of particular facts, stripped of the poetry which should invest them, augments that of poetry, and for ever develops new and wonderful applications of the eternal truth which it contains. Hence epitomes have been called the moths of just history ; they eat out the poetry of it. A story of particular facts is as a mirror which obscures and distorts that which should be beautiful : poetry is a mirror which makes beautiful that which is distorted.

The parts of a composition may be poetical, without the composition as a whole being a poem. A single sentence may be considered as a whole, though it may be found in the midst of a series of unassimilated portions : a single word even may be a spark of inextinguishable thought. And thus all the great historians, Herodotus, Plutarch, Livy, were poets ; and although the plan of these writers, especially that of Livy, restrained them from developing this faculty in its highest degree, they made copious and ample amends for their subjection, by filling all the interstices of their subjects with living images.

Having determined what is poetry, and who are poets, let us proceed to estimate its effects upon society.

Poetry is ever accompanied with pleasure: all spirits on which it falls open themselves to receive the wisdom which is mingled with its delight. In the infancy of the world, neither poets themselves nor their auditors are fully aware of the excellence of poetry: for it acts in a divine and unapprehended manner, beyond and above consciousness; and it is reserved for future generations to contemplate and measure the mighty cause and effect in all the strength and splendour of their union. Even in modern times, no living poet ever arrived at the fullness of his fame; the jury which sits in judgement upon a poet, belonging as he does to all time, must be composed of his peers: it must be impanelled by Time from the selectest of the wise of many generations. A poet is a nightingale, who sits in darkness and sings to cheer its own solitude with sweet sounds; his auditors are as men entranced by the melody of an unseen musician, who feel that they are moved and softened, yet know not whence or why. The poems of Homer and his contemporaries were the delight of infant Greece; they were the elements of that social system which is the column upon which all succeeding civilization has reposed. Homer embodied the ideal perfection of his age in human character; nor can we doubt that those who read his verses were awakened to an ambition of becoming like to Achilles, Hector, and Ulysses: the truth and beauty of friendship, patriotism, and persevering devotion to an object, were unveiled to the depths in these immortal creations: the sentiments of the auditors must have been refined and enlarged by a sympathy with such great and lovely impersonations, until from admiring they imitated, and from imitation they identified themselves with the objects of their admiration. Nor let it be objected, that these characters are remote from moral perfection, and that they can by no means be considered as edifying patterns for general imitation. Every epoch, under names more or less specious, has deified its peculiar errors; Revenge is the naked idol of the worship of a semi-barbarous age; and Self-deceit is the veiled image of unknown evil, before which luxury and satiety lie prostrate. But a poet considers the vices of his contemporaries as a temporary dress in which his creations must be arrayed, and which cover without concealing the eternal proportions of their beauty. An epic or dramatic personage is understood to wear them around his soul, as he may the ancient armour or the modern uniform around his body; whilst it is easy to conceive a dress more graceful than either. The beauty of the internal nature cannot be so far concealed by its accidental vesture, but that the spirit of its form shall communicate itself to the very disguise, and indicate the shape it hides from the manner in which it is worn. A majestic form and graceful motions will express themselves through the most barbarous and tasteless costume. Few poets of the highest class have chosen to exhibit the beauty of their conceptions in its naked truth and splendour; and it is doubtful whether the alloy of costume, habit, &c., be not necessary to temper this planetary music for mortal ears.

The whole objection, however, of the immorality of poetry rests upon a misconception of the manner in which poetry acts to produce the moral improvement of man. Ethical science arranges the elements which poetry has created, and propounds schemes and proposes examples of civil and domestic life: nor is it for want of admirable doctrines that men hate, and despise, and censure, and deceive, and subjugate one another. But poetry acts in another and diviner manner. It awakens and enlarges the mind itself by rendering it the receptacle of a thousand unapprehended combinations of thought. Poetry lifts the veil from the hidden beauty of the world, and makes familiar objects be as if they were not familiar; it reproduces all that it represents, and the impersonations clothed in its Elysian light stand thenceforward in the minds of those who have once contemplated them, as memorials of that gentle and exalted content which extends itself over all thoughts and actions with which it coexists. The great secret of morals is love; or a going out of our own nature, and an identification of ourselves with the beautiful which exists in thought, action, or person, not our own. A man, to be greatly good, must imagine intensely and comprehensively; he must put himself in the place of another and of many others; the pains and pleasures of his species must become his own. The great instrument of moral good is the imagination; and poetry administers to the effect by acting upon the cause. Poetry enlarges the circumference of the imagination by replenishing it with thoughts of ever new delight, which have the power of attracting and assimilating to their own nature all other thoughts, and which form new intervals and interstices whose void for ever craves fresh food. Poetry strengthens the faculty which is the organ of the moral nature of man, in the same manner as exercise strengthens a limb. A poet therefore would do ill to embody his own conceptions of right and wrong, which are usually those of his place and time, in his poetical creations, which participate in neither. By this assumption of the inferior office of interpreting the effect, in which perhaps after all he might acquit himself but imperfectly, he would resign a glory in a participation in the cause. There was little danger that Homer, or any of the eternal poets, should have so far misunderstood themselves as to have abdicated this throne of their widest dominion. Those in whom the poetical faculty, though great, is less intense, as Euripides, Lucan, Tasso, Spenser, have frequently affected a moral aim, and the effect of their poetry is diminished in exact proportion to the degree in which they compel us to advert to this purpose.

Homer and the cyclic poets were followed at a certain interval by the dramatic and lyrical poets of Athens, who flourished contemporaneously with all that is most perfect in the kindred expressions of the poetical faculty; architecture, painting, music, the dance, sculpture, philosophy, and, we may add, the forms of civil life. For although the scheme of Athenian society was deformed by many imperfections which the poetry existing in chivalry and Christianity has erased from the habits and institutions of modern Europe; yet never at any other period has so much energy, beauty, and virtue, been developed; never was blind strength and stubborn form so disciplined and rendered subject to the will of man, or that will less repugnant to the dictates of the beautiful and the true, as during the century which preceded the death of Socrates. Of no other epoch in the history of our species have we records and fragments stamped so visibly with the image of the divinity in man. But it is poetry alone, in form, in action, or in language, which has rendered this epoch memorable above all others, and the storehouse of examples to everlasting time. For written poetry existed at that epoch simultaneously with the other arts, and it is an idle inquiry to demand which gave and which received the light, which all, as from a common focus, have scattered over the darkest periods of succeeding time. We know no more of cause and effect than a constant conjunction of events: poetry is ever found to co-exist with whatever other arts contribute to the happiness and perfection of man. I appeal to what has already been established to distinguish between the cause and the effect.

It was at the period here adverted to, that the drama had its birth; and however a succeeding writer may have equalled or surpassed those few great specimens of the Athenian drama which have been preserved to us, it is indisputable that the art itself never was understood or practised according to the true philosophy of it, as at Athens. For the Athenians employed language, action, music, painting, the dance, and religious institutions, to produce a common effect in the representation of the highest idealisms of passion and of power; each division in the art was made perfect in its kind by artists of the most consummate skill, and was disciplined into a beautiful proportion and unity one towards the other. On the modern stage a few only of the elements capable of expressing the image of the poet's conception are employed at once. We have tragedy without music and dancing; and music and dancing without the highest impersonations of which they are the fit accompaniment, and both without religion and solemnity. Religious institution has indeed been usually banished from the stage. Our system of divesting the actor's face of a mask, on which the many expressions appropriated to his dramatic character might be moulded into one permanent and unchanging expression, is favourable only to a partial and inharmonious effect; it is fit for nothing but a monologue, where all the attention may be directed to some great master of ideal mimicry. The modern practice of blending comedy with tragedy, though liable to great abuse in point of practice, is undoubtedly an extension of the dramatic circle; but the comedy should be as in *King Lear*, universal, ideal, and sublime. It is perhaps the intervention of this principle which determines the balance in favour of *King Lear* against the *Oedipus Tyrannus* or the *Agamemnon*, or, if you will, the trilogies with which they are connected; unless the intense power of the choral poetry, especially that of the latter, should be considered as restoring the equilibrium. *King Lear*, if it can sustain this comparison, may be judged to be the most perfect specimen of the dramatic art existing in the world; in spite of the narrow conditions to which the poet was subjected by the ignorance of the philosophy of the drama which has prevailed in modern Europe. Calderon, in his religious *Autos*, has attempted to fulfil some of the high conditions of dramatic representation neglected by Shakespeare; such as the establishing a relation between the drama and religion, and the accommodating them to music and dancing; but he omits the observation of conditions still more important, and more is lost than gained by the substitution of the rigidly-defined and ever-repeated idealisms of a distorted superstition for the living impersonations of the truth of human passion.

But I digress.—The connexion of scenic exhibitions with the improvement or corruption of the manners of men, has been universally recognized: in other words, the presence or absence of poetry in its most perfect and universal form, has been found to be connected with good and evil in conduct or habit. The corruption which has been imputed to the drama as an effect, begins, when the poetry employed in its constitution ends: I appeal to the history of manners whether the periods of the growth of the one and the decline of the other have not corresponded with an exactness equal to any example of moral cause and effect.

The drama at Athens, or wheresoever else it may have approached to its perfection, ever co-existed with the moral and intellectual greatness of the age. The tragedies of the Athenian poets are as mirrors in which the spectator beholds himself, under a thin disguise of circumstance, stript of all but that ideal perfection and energy which every one feels to be the internal type of all that he loves, admires, and would become. The imagination is enlarged by a sympathy with pains and passions so mighty, that they distend in their conception the capacity of that by which they are conceived; the good affections are strengthened by pity, indignation, terror, and sorrow; and an exalted calm is prolonged from the satiety of this high exercise of them into the tumult of familiar life: even crime is disarmed of half its horror and all its contagion by being represented as the fatal consequence of the unfathomable agencies of nature; error is thus divested of its wilfulness; men can no longer cherish it as the creation of their choice. In a drama of the highest order there is little food for censure or hatred; it teaches rather self-knowledge and self-respect. Neither the eye nor the mind can see itself, unless reflected upon that which it resembles. The drama, so long as it continues to express poetry, is as a prismatic and many-sided mirror, which collects the brightest rays of human nature and divides and reproduces them from the simplicity of these elementary forms, and touches them with majesty and beauty, and multiplies all that it reflects, and endows it with the power of propagating its like wherever it may fall.

But in periods of the decay of social life, the drama sympathizes with that decay. Tragedy becomes a cold imitation of the form of the great masterpieces of antiquity, divested of all harmonious accompaniment of the kindred arts; and often the very form misunderstood, or a weak attempt to teach certain doctrines, which the writer considers as moral truths; and which are usually no more than specious flatteries of some gross vice or weakness, with which the author, in common with his auditors, are infected. Hence what has been called the classical and domestic drama. Addison's *Cato* is a specimen of the one; and would it were not superfluous to cite examples of the other! To such purposes poetry cannot be made subservient. Poetry is a sword of lightning, ever unsheathed, which consumes the scabbard that would contain it. And thus we observe that all dramatic writings of this nature are unimaginative in a singular degree; they affect sentiment and passion, which, divested of imagination, are other names for caprice and appetite. The period in our own history of the grossest degradation of the drama is the reign of Charles II, when all forms in which poetry had been accustomed to be expressed became hymns to the triumph of kingly power over liberty and virtue. Milton stood alone illuminating an age unworthy of him. At such periods the calculating principle pervades all the forms of dramatic exhibition, and poetry ceases to be expressed upon them. Comedy loses its ideal universality: wit succeeds to humour; we laugh from self-complacency and triumph, instead of pleasure; malignity, sarcasm, and contempt, succeed to sympathetic merriment; we hardly laugh, but we smile. Obscenity, which is ever blasphemy against the divine beauty in life, becomes, from the very veil which it assumes, more active if less disgusting : it is a monster for which the corruption of society for ever brings forth new food, which it devours in secret.

The drama being that form under which a greater number of modes of expression of poetry are susceptible of being combined than any other, the connexion of poetry and social good is more observable in the drama than in whatever other form. And it is indisputable that the highest perfection of human society has ever corresponded with the highest dramatic excellence ; and that the corruption or the extinction of the drama in a nation where it has once flourished, is a mark of a corruption of manners, and an extinction of the energies which sustain the soul of social life. But, as Machiavelli says of political institutions, that life may be preserved and renewed, if men should arise capable of bringing back the drama to its principles. And this is true with respect to poetry in its most extended sense : all language, institution and form, require not only to be produced but to be sustained : the office and character of a poet participates in the divine nature as regards providence, no less than as regards creation.

Civil war, the spoils of Asia, and the fatal predominance first of the Macedonian, and then of the Roman arms, were so many symbols of the extinction or suspension of the creative faculty in Greece. The bucolic writers, who found patronage under the lettered tyrants of Sicily and Egypt, were the latest representatives of its most glorious reign. Their poetry is intensely melodious ; like the odour of the tuberose, it overcomes and sickens the spirit with excess of sweetness ; whilst the poetry of the preceding age was as a meadow-gale of June, which mingles the fragrance of all the flowers of the field, and adds a quickening and harmonizing spirit of its own, which endows the sense with a power of sustaining its extreme delight. The bucolic and erotic delicacy in written poetry is correlative with that softness in statuary, music, and the kindred arts, and even in manners and institutions, which distinguished the epoch to which I now refer. Nor is it the poetical faculty itself, or any misapplication of it, to which this want of harmony is to be imputed. An equal sensibility to the influence of the senses and the affections is to be found in the writings of Homer and Sophocles : the former, especially, has clothed sensual and pathetic images with irresistible attractions. Their superiority over these succeeding writers consists in the presence of those thoughts which belong to the inner faculties of our nature, not in the absence of those which are connected with the external : their incomparable perfection consists in a harmony of the union of all. It is not what the erotic poets have, but what they have not, in which their imperfection consists. It is not inasmuch as they were poets, but inasmuch as they were not poets, that they can be considered with any plausibility as connected with the corruption of their age. Had that corruption availed so as to extinguish in them the sensibility to pleasure, passion, and natural scenery, which is imputed to them as an imperfection, the last triumph of evil would have been achieved. For the end of social corruption is to destroy all sensibility to pleasure ; and, therefore, it is corruption. It begins at the imagination and the intellect as at the core, and distributes itself thence as a paralysing venom, through the affections into the very appetites, until all become a torpid mass in which hardly sense survives. At the approach of such a period, poetry ever addresses itself to those faculties which are the last to be destroyed, and its voice is heard, like the footsteps of Astraea, departing from the world. Poetry ever communicates all the pleasure which men are capable of receiving: it is ever still the light of life; the source of whatever of beautiful or generous or true can have place in an evil time. It will readily be confessed that those among the luxurious citizens of Syracuse and Alexandria, who were delighted with the poems of Theocritus, were less cold, cruel, and sensual than the remnant of their tribe. But corruption must utterly have destroyed the fabric of human society before poetry can ever cease. The sacred links of that chain have never been entirely disjoined, which descending through the minds of many men is attached to those great minds, whence as from a magnet the invisible effluence is sent forth, which at once connects, animates, and sustains the life of all. It is the faculty which contains within itself the seeds at once of its own and of social renovation. And let us not circumscribe the effects of the bucolic and erotic poetry within the limits of the sensibility of those to whom it was addressed. They may have perceived the beauty of those immortal compositions, simply as fragments and isolated portions: those who are more finely organized, or born in a happier age, may recognize them as episodes to that great poem, which all poets, like the co-operating thoughts of one great mind, have built up since the beginning of the world.

The same revolutions within a narrower sphere had place in ancient Rome; but the actions and forms of its social life never seem to have been perfectly saturated with the poetical element. The Romans appear to have considered the Greeks as the selectest treasuries of the selectest forms of manners and of nature, and to have abstained from creating in measured language, sculpture, music, or architecture, anything which might bear a particular relation to their own condition, whilst it should bear a general one to the universal constitution of the world. But we judge from partial evidence, and we judge perhaps partially. Ennius, Varro, Pacuvius, and Accius, all great poets, have been lost. Lucretius is in the highest, and Virgil in a very high sense, a creator. The chosen delicacy of expressions of the latter, are as a mist of light which conceal from us the intense and exceeding truth of his conceptions of nature. Livy is instinct with poetry. Yet Horace, Catullus, Ovid, and generally the other great writers of the Virgilian age, saw man and nature in the mirror of Greece. The institutions also, and the religion of Rome were less poetical than those of Greece, as the shadow is less vivid than the substance. Hence poetry in Rome, seemed to follow, rather than accompany, the perfection of political and domestic society. The true poetry of Rome lived in its institutions; for whatever of beautiful, true, and majestic, they contained, could have sprung only from the faculty which creates the order in which they consist. The life of Camillus, the death of Regulus; the expectation of the senators, in their godlike state, of the victorious Gauls: the refusal of the republic to make peace with Hannibal, after the battle of Cannae, were not the consequences of a refined calculation of the probable personal advantage to result from such a rhythm and order in the shows of life, to those who were at once the poets and the actors of these immortal dramas. The imagination beholding the beauty of this order, created it out of itself according to its own idea; the consequence was empire, and the reward everliving fame. These things are not the less poetry *quia carent vate sacro*. They are the episodes of that cyclic poem written by Time upon the memories of men. The Past, like an inspired rhapsodist, fills the theatre of everlasting generations with their harmony.

At length the ancient system of religion and manners had fulfilled the circle of its revolutions. And the world would have fallen into utter anarchy and darkness, but that there were found poets among the authors of the Christian and chivalric systems of manners and relig on, who created forms of opinion and action never before conceived; which, copied into the imaginations of men, become as generals to the bewildered armies of their thoughts. It is foreign to the present purpose to touch upon the evil produced by these systems: except that we protest, on the ground of the principles already established, that no portion of it can be attributed to the poetry they contain.

It is probable that the poetry of Moses, Job, David, Solomon, and Isaiah, had produced a great effect upon the mind of Jesus and his disciples. The scattered fragments preserved to us by the biographers of this extraordinary person, are all instinct with the most vivid poetry. But his doctrines seem to have been quickly distorted. At a certain period after the prevalence of a system of opinions founded upon those promulgated by him, the three forms into which Plato had distributed the faculties of mind underwent a sort of apotheosis, and became the object of the worship of the civilized world. Here it is to be confessed that 'Light seems to thicken', and

> The crow makes wing to the rooky wood,
> Good things of day begin to droop and drowse,
> And night's black agents to their preys do rouze.

But mark how beautiful an order has sprung from the dust and blood of this fierce chaos! how the world, as from a resurrection, balancing itself on the golden wings of knowledge and of hope, has reassumed its yet unwearied flight into the heaven of time. Listen to the music, unheard by outward ears, which is as a ceaseless and invisible wind, nourishing its everlasting course with strength and swiftness.

The poetry in the doctrines of Jesus Christ, and the mythology and institutions of the Celtic conquerors of the Roman empire, outlived the darkness and the convulsions connected with their growth and victory, and blended themselves in a new fabric of manners and opinion. It is an error to impute the ignorance of the dark ages to the Christian doctrines or the predominance of the Celtic nations. Whatever of evil their agencies may have contained sprang from the extinction of the poetical principle, connected with the progress of despotism and superstition. Men, from causes too intricate to be here discussed, had become insensible and selfish : their own will had become feeble, and yet they were its slaves, and thence the slaves of the will of others : lust, fear, avarice, cruelty, and fraud, characterized a race amongst whom no one was to be found capable of *creating* in form, language, or institution. The moral anomalies of such a state of society are not justly to be charged upon any class of events immediately connected with them, and those events are most entitled to our approbation which could dissolve it most expeditiously. It is unfortunate for those who cannot distinguish words from thoughts, that many of these anomalies have been incorporated into our popular religion.

It was not until the eleventh century that the effects of the poetry of the Christian and chivalric systems began to manifest themselves. The principle of equality had been discovered and applied by Plato in his *Republic*, as the theoretical rule of the mode in which the materials of pleasure and of power, produced by the common skill and labour of human beings, ought to be distributed among them. The limitations of this rule were asserted by him to be determined only by the sensibility of each, or the utility to result to all. Plato, following the doctrines of Timaeus and Pythagoras, taught also a moral and intellectual system of doctrine, comprehending at once the past, the present, and the future condition of man. Jesus Christ divulged the sacred and eternal truths contained in these views to mankind, and Christianity, in its abstract purity, became the exoteric expression of the esoteric doctrines of the poetry and wisdom of antiquity. The incorporation of the Celtic nations with the exhausted population of the south, impressed upon it the figure of the poetry existing in their mythology and institutions. The result was a sum of the action and reaction of all the causes included in it; for it may be assumed as a maxim that no nation or religion can supersede any other without incorporating into itself a portion of that which it supersedes. The abolition of personal and domestic slavery, and the emancipation of women from a great part of the degrading restraints of antiquity, were among the consequences of these events.

The abolition of personal slavery is the basis of the highest political hope that it can enter into the mind of man to conceive. The freedom of women produced the poetry of sexual love. Love became a religion, the idols of whose worship were ever present. It was as if the statues of Apollo and the Muses had been endowed with life and motion, and had walked forth among their worshippers; so that earth became peopled by the inhabitants of a diviner world. The familiar appearance and proceedings of life became wonderful and heavenly, and a paradise was created as out of the wrecks of Eden. And as this creation itself is poetry, so its creators were poets; and language was the instrument of their art: 'Galeotto fù il libro, e chi lo scrisse.' The Provençal Trouveurs, or inventors, preceded Petrarch, whose verses are as spells, which unseal the inmost enchanted fountains of the delight which is in the grief of love. It is impossible to feel them without becoming a portion of that beauty which we contemplate: it were superfluous to explain how the gentleness and the elevation of mind connected with these sacred emotions can render men more amiable, more generous and wise, and lift them out of the dull vapours of the little world of self. Dante understood the secret things of love even more than Petrarch. His *Vita Nuova* is an inexhaustible fountain of purity of sentiment and language: it is the idealized history of that period, and those intervals of his life which were dedicated to love. His apotheosis of Beatrice in Paradise, and the gradations of his own love and her loveliness, by which as by steps he feigns himself to have ascended to the throne of the Supreme Cause, is the most glorious imagination of modern poetry. The acutest critics have justly reversed the judgement of the vulgar, and the order of the great acts of the 'Divine Drama', in the measure of the admiration which they accord to the Hell, Purgatory, and Paradise. The latter is a perpetual hymn of everlasting love. Love, which found a worthy poet in Plato alone of all the ancients, has been celebrated by a chorus of the greatest writers of the renovated world; and the music has penetrated the caverns of society, and its echoes still drown the dissonance of arms and superstition. At successive intervals, Ariosto, Tasso, Shakespeare, Spenser, Calderon, Rousseau, and the great writers of our own age, have celebrated the dominion of love, planting as it were trophies in the human mind of that sublimest victory over sensuality and force. The true relation borne to each other by the sexes into which human kind is distributed, has become less misunderstood; and if the error which confounded diversity with inequality of the powers of the two sexes has been partially recognized in the opinions and institutions of modern Europe, we owe this great benefit to the worship of which chivalry was the law, and poets the prophets.

The poetry of Dante may be considered as the bridge thrown over the stream of time, which unites the modern and ancient world. The distorted notions of invisible things which Dante and his rival Milton have idealized, are merely the mask and the mantle in which these great poets walk through eternity enveloped and disguised. It is a difficult question to determine how far they were conscious of the distinction which must have subsisted in their minds between their own creeds and that of the people. Dante at least appears to wish to mark the full extent of it by placing Riphaeus, whom Virgil calls *justissimus unus*, in Paradise, and observing a most heretical caprice in his distribution of rewards and punishments. And Milton's poem contains within itself a philosophical refutation of that system, of which, by a strange and natural antithesis, it has been a chief popular support. Nothing can exceed the energy and magnificence of the character of Satan as expressed in *Paradise Lost*. It is a mistake to suppose that he could ever have been intended for the popular personification of evil. Implacable hate, patient cunning, and a sleepless refinement of device to inflict the extremest anguish on an enemy, these things are evil; and, although venial in a slave, are not to be forgiven in a tyrant; although redeemed by much that ennobles his defeat in one subdued, are marked by all that dishonours his conquest in the victor. Milton's Devil as a moral being is as far superior to his God, as one who perseveres in some purpose which he has conceived to be excellent in spite of adversity and torture, is to one who in the cold security of undoubted triumph inflicts the most horrible revenge upon his enemy, not from any mistaken notion of inducing him to repent of a perseverance in enmity, but with the alleged design of exasperating him to deserve new torments. Milton has so far violated the popular creed (if this shall be judged to be a violation) as to have alleged no superiority of moral virtue to his God over his Devil. And this bold neglect of a direct moral purpose is the most decisive proof of the supremacy of Milton's genius. He mingled as it were the elements of human nature as colours upon a single pallet, and arranged them in the composition of his great picture according to the laws of epic truth; that is, according to the laws of that principle by which a series of actions of the external universe and of intelligent and ethical beings is calculated to excite the sympathy of succeeding generations of mankind. The *Divina Commedia* and *Paradise Lost* have conferred upon modern mythology a systematic form; and when change and time shall have added one more superstition to the mass of those which have arisen and decayed upon the earth, commentators will be learnedly employed in elucidating the religion of ancestral Europe, only not utterly forgotten because it will have been stamped with the eternity of genius.

Homer was the first and Dante the second epic poet: that is, the second poet, the series of whose creations bore a defined and intelligible relation to the knowledge and sentiment and religion of the age in which he lived, and of the ages which followed it:

developing itself in correspondence with their development. For Lucretius had limed the wings of his swift spirit in the dregs of the sensible world; and Virgil, with a modesty that ill became his genius, had affected the fame of an imitator, even whilst he created anew all that he copied; and none among the flock of mock-birds, though their notes were sweet, Apollonius Rhodius, Quintus Calaber, Nonnus, Lucan, Statius, or Claudian, have sought even to fulfil a single condition of epic truth. Milton was the third epic poet. For if the title of epic in its highest sense be refused to the *Aeneid*, still less can it be conceded to the *Orlando Furioso*, the *Gerusalemme Liberata*, the *Lusiad*, or the *Fairy Queen*.

Dante and Milton were both deeply penetrated with the ancient religion of the civilized world; and its spirit exists in their poetry probably in the same proportion as its forms survived in the unreformed worship of modern Europe. The one preceded and the other followed the Reformation at almost equal intervals. Dante was the first religious reformer, and Luther surpassed him rather in the rudeness and acrimony, than in the boldness of his censures of papal usurpation. Dante was the first awakener of entranced Europe; he created a language, in itself music and persuasion, out of a chaos of inharmonious barbarisms. He was the congregator of those great spirits who presided over the resurrection of learning; the Lucifer of that starry flock which in the thirteenth century shone forth from republican Italy, as from a heaven, into the darkness of the benighted world. His very words are instinct with spirit; each is as a spark, a burning atom of inextinguishable thought; and many yet lie covered in the ashes of their birth, and pregnant with a lightning which has yet found no conductor. All high poetry is infinite; it is as the first acorn, which contained all oaks potentially. Veil after veil may be undrawn, and the inmost naked beauty of the meaning never exposed. A great poem is a fountain for ever overflowing with the waters of wisdom and delight; and after one person and one age has exhausted all its divine effluence which their peculiar relations enable them to share, another and yet another succeeds, and new relations are ever developed, the source of an unforeseen and an unconceived delight.

The age immediately succeeding to that of Dante, Petrarch, and Boccaccio, was characterized by a revival of painting, sculpture, and architecture. Chaucer caught the sacred inspiration, and the superstructure of English literature is based upon the materials of Italian invention.

But let us not be betrayed from a defence into a critical history of poetry and its influence on society. Be it enough to have pointed out the effects of poets, in the large and true sense of the word, upon their own and all succeeding times.

But poets have been challenged to resign the civic crown to reasoners and mechanists, on another plea. It is admitted that the exercise of the imagination is most delightful, but it is alleged that that of reason is more useful. Let us examine as the grounds of this distinction, what is here meant by utility. Pleasure or good, in a general sense, is that which the consciousness of a sensitive and intelligent being seeks, and in which, when found, it acquiesces. There are two kinds of pleasure, one durable, universal and permanent; the other transitory and particular. Utility may either express the means of producing the former or the latter. In the former sense, whatever strengthens and purifies the affections, enlarges the imagination, and adds spirit to sense, is useful. But a narrower meaning may be assigned to the word utility, confining it to express that which banishes the importunity of the wants of our animal nature, the surrounding men with security of life, the dispersing the grosser delusions of superstition, and the conciliating such a degree of mutual forbearance among men as may consist with the motives of personal advantage.

Undoubtedly the promoters of utility, in this limited sense, have their appointed office in society. They follow the footsteps of poets, and copy the sketches of their creations into the book of common life. They make space, and give time. Their exertions are of the highest value, so long as they confine their administration of the concerns of the inferior powers of our nature within the limits due to the superior ones. But whilst the sceptic destroys gross superstitions, let him spare to deface, as some of the French writers have defaced, the eternal truths charactered upon the imaginations of men. Whilst the mechanist abridges, and the political economist combines labour, let them beware that their speculations, for want of correspondence with those first principles which belong to the imagination, do not tend, as they have in modern England, to exasperate at once the extremes of luxury and want. They have exemplified the saying, 'To him that hath, more shall be given; and from him that hath not, the little that he hath shall be taken away.' The rich have become richer, and the poor have become poorer; and the vessel of the state is driven between the Scylla and Charybdis of anarchy and despotism. Such are the effects which must ever flow from an unmitigated exercise of the calculating faculty.

It is difficult to define pleasure in its highest sense; the definition involving a number of apparent paradoxes. For, from an inexplicable defect of harmony in the constitution of human nature, the pain of the inferior is frequently connected with the pleasures of the superior portions of our being. Sorrow, terror, anguish, despair itself, are often the chosen expressions of an approximation to the highest good. Our sympathy in tragic fiction depends on this principle; tragedy delights by affording a shadow of the pleasure which exists in pain. This is the source also of the melancholy which is inseparable from the sweetest melody. The pleasure that is in sorrow is sweeter than the pleasure of pleasure itself. And hence the saying, 'It is better to go to the house of mourning, than to the house of mirth.' Not that this highest species of pleasure is necessarily linked with pain. The delight of love and friendship, the ecstasy of the admiration of nature, the joy of the perception and still more of the creation of poetry, is often wholly unalloyed.

The production and assurance of pleasure in this highest sense is true utility. Those who produce and preserve this pleasure are poets or poetical philosophers.

The exertions of Locke, Hume, Gibbon, Voltaire, Rousseau,[1] and their disciples, in favour of oppressed and deluded humanity, are entitled to the gratitude of mankind. Yet it is easy to calculate the degree of moral and intellectual improvement which the world would have exhibited, had they never lived. A little more nonsense would have been talked for a century or two; and perhaps a few more men, women, and children, burnt as heretics. We might not at this moment have been congratulating each other on the abolition of the Inquisition in Spain. But it exceeds all imagination to conceive what would have been the moral condition of the world if neither Dante, Petrarch,

[1] Although Rousseau has been thus classed, he was essentially a poet. The others, even Voltaire, were mere reasoners.

Boccaccio, Chaucer, Shakespeare, Calderon, Lord Bacon, nor Milton, had ever existed ; if Raphael and Michael Angelo had never been born ; if the Hebrew poetry had never been translated ; if a revival of the study of Greek literature had never taken place ; if no monuments of ancient sculpture had been handed down to us ; and if the poetry of the religion of the ancient world had been extinguished together with its belief. The human mind could never, except by the intervention of these excitements, have been awakened to the invention of the grosser sciences, and that application of analytical reasoning to the aberrations of society, which it is now attempted to exalt over the direct expression of the inventive and creative faculty itself.

We have more moral, political and historical wisdom, than we know how to reduce into practice ; we have more scientific and economical knowledge than can be accommodated to the just distribution of the produce which it multiplies. The poetry in these systems of thought, is concealed by the accumulation of facts and calculating processes. There is no want of knowledge respecting what is wisest and best in morals, government, and political economy, or at least, what is wiser and better than what men now practise and endure. But we let '*I dare not* wait upon *I would*, like the poor cat in the adage.' We want the creative faculty to imagine that which we know ; we want the generous impulse to act that which we imagine ; we want the poetry of life : our calculations have outrun conception ; we have eaten more than we can digest. The cultivation of those sciences which have enlarged the limits of the empire of man over the external world, has, for want of the poetical faculty, proportionally circumscribed those of the internal world ; and man, having enslaved the elements, remains himself a slave. To what but a cultivation of the mechanical arts in a degree disproportioned to the presence of the creative faculty, which is the basis of all knowledge, is to be attributed the abuse of all invention for abridging and combining labour, to the exasperation of the inequality of mankind ? From what other cause has it arisen that the discoveries which should have lightened, have added a weight to the curse imposed on Adam ? Poetry, and the principle of Self, of which money is the visible incarnation, are the God and Mammon of the world.

The functions of the poetical faculty are two-fold ; by one it creates new materials of knowledge and power and pleasure ; by the other it engenders in the mind a desire to reproduce and arrange them according to a certain rhythm and order which may be called the beautiful and the good. The cultivation of poetry is never more to be desired than at periods when, from an excess of the selfish and calculating principle, the accumulation of the materials of external life exceed the quantity of the power of assimilating them to the internal laws of human nature. The body has then become too unwieldy for that which animates it.

Poetry is indeed something divine. It is at once the centre and circumference of knowledge ; it is that which comprehends all science, and that to which all science must be referred. It is at the same time the root and blossom of all other systems of thought ; it is that from which all spring, and that which adorns all ; and that which, if blighted, denies the fruit and the seed, and withholds from the barren world the nourishment and the succession of the scions of the tree of life. It is the perfect and consummate surface and bloom of all things ; it is as the odour and the colour of the rose to the texture of the elements which compose it, as the form and splendour of unfaded beauty to the secrets of anatomy and corruption. What were virtue, love, patriotism, friendship—what were the scenery of this beautiful universe which we inhabit; what were our consolations on this side of the grave—and what were our aspirations beyond it, if poetry did not ascend to bring light and fire from those eternal regions where the owl-winged faculty of calculation dare not ever soar? Poetry is not like reasoning, a power to be exerted according to the determination of the will. A man cannot say, 'I will compose poetry.' The greatest poet even cannot say it; for the mind in creation is as a fading coal, which some invisible influence, like an inconstant wind, awakens to transitory brightness; this power arises from within, like the colour of a flower which fades and changes as it is developed, and the conscious portions of our natures are unprophetic either of its approach or its departure. Could this influence be durable in its original purity and force, it is impossible to predict the greatness of the results; but when composition begins, inspiration is already on the decline, and the most glorious poetry that has ever been communicated to the world is probably a feeble shadow of the original conceptions of the poet. I appeal to the greatest poets of the present day, whether it is not an error to assert that the finest passages of poetry are produced by labour and study. The toil and the delay recommended by critics, can be justly interpreted to mean no more than a careful observation of the inspired moments, and an artificial connexion of the spaces between their suggestions by the intertexture of conventional expressions; a necessity only imposed by the limitedness of the poetical faculty itself; for Milton conceived the *Paradise Lost* as a whole before he executed it in portions. We have his own authority also for the muse having 'dictated' to him the 'unpremeditated song'. And let this be an answer to those who would allege the fifty-six various readings of the first line of the *Orlando Furioso.* Compositions so produced are to poetry what mosaic is to painting. This instinct and intuition of the poetical faculty is still more observable in the plastic and pictorial arts; a great statue or picture grows under the power of the artist as a child in the mother's womb; and the very mind which directs the hands in formation is incapable of accounting to itself for the origin, the gradations, or the media of the process.

Poetry is the record of the best and happiest moments of the happiest and best minds. We are aware of evanescent visitations of thought and feeling sometimes associated with place or person, sometimes regarding our own mind alone, and always arising unforeseen and departing unbidden, but elevating and delightful beyond all expression: so that even in the desire and regret they leave, there cannot but be pleasure, participating as it does in the nature of its object. It is as it were the interpenetration of a diviner nature through our own; but its footsteps are like those of a wind over the sea, which the coming calm erases, and whose traces remain only, as on the wrinkled sand which paves it. These and corresponding conditions of being are experienced principally by those of the most delicate sensibility and the most enlarged imagination; and the state of mind produced by them is at war with every base desire. The enthusiasm of virtue, love, patriotism, and friendship, is essentially linked with such emotions; and whilst they last, self appears as what it is, an atom to a universe. Poets are not only subject to these experiences as spirits of the most refined organization, but they can colour all that they combine with the evanescent hues of this ethereal world; a word, a trait in the representation of a scene or a passion, will touch the enchanted chord, and reanimate, in those who have ever experienced these emotions, the sleeping, the cold, the buried image of the past. Poetry thus makes immortal all that is best and most beautiful in the world; it arrests the vanishing apparitions which haunt the interlunations of life, and veiling them, or in language or in form, sends them forth among mankind, bearing sweet news of kindred joy to those with whom their sisters abide—abide, because there is no portal of expression from the caverns of the spirit which they inhabit into the universe of things. Poetry redeems from decay the visitations of the divinity in man.

Poetry turns all things to loveliness; it exalts the beauty of that which is most beautiful, and it adds beauty to that which is most deformed; it marries exultation and horror, grief and pleasure, eternity and change; it subdues to union under its light yoke, all irreconcilable things. It transmutes all that it touches, and every form moving within the radiance of its presence is changed by wondrous sympathy to an incarnation of the spirit which it breathes: its secret alchemy turns to potable gold the poisonous waters which flow from death through life; it strips the veil of familiarity from the world, and lays bare the naked and sleeping beauty, which is the spirit of its forms.

All things exist as they are perceived; at least in relation to the percipient. 'The mind is its own place, and of itself can make a heaven of hell, a hell of heaven.' But poetry defeats the curse which binds us to be subjected to the accident of surrounding impressions. And whether it spreads its own figured curtain, or withdraws life's dark veil from before the scene of things, it equally creates for us a being within our being. It makes us the inhabitants of a world to which the familiar world is a chaos. It reproduces the common universe of which we are portions and percipients, and it purges from our inward sight the film of familiarity which obscures from us the wonder of our being. It compels us to feel that which we perceive, and to imagine that which we know. It creates anew the universe, after it has been annihilated in our minds by the recurrence of impressions blunted by reiteration. It justifies the bold and true words of Tasso : *Non merita nome di creatore, se non Iddio ed il Poeta.*

A poet, as he is the author to others of the highest wisdom, pleasure, virtue and glory, so he ought personally to be the happiest, the best, the wisest, and the most illustrious of men. As to his glory, let time be challenged to declare whether the fame of any other institutor of human life be comparable to that of a poet. That he is the wisest, the happiest, and the best, inasmuch as he is a poet, is equally incontrovertible : the greatest poets have been men of the most spotless virtue, of the most consummate prudence, and, if we would look into the interior of their lives, the most fortunate of men : and the exceptions, as they regard those who possessed the poetic faculty in a high yet inferior degree, will be found on consideration to confine rather than destroy the rule. Let us for a moment stoop to the arbitration of popular breath, and usurping and uniting in our own persons the incompatible characters of accuser, witness, judge, and executioner, let us decide without trial, testimony, or form, that certain motives of those who are 'there sitting where we dare not soar', are reprehensible. Let us assume that Homer was a drunkard, that Virgil was a flatterer, that Horace was a coward, that Tasso was a madman, that Lord Bacon was a peculator, that Raphael was a libertine, that Spenser was a poet laureate. It is inconsistent with this division of our subject to cite living poets, but posterity has done ample justice to the great names now referred to. Their errors have been weighed and found to have been dust in the balance; if their sins 'were as scarlet, they are now white as snow': they have been washed in the blood of the mediator and redeemer, Time. Observe in what a ludicrous chaos the imputations of real or fictitious crime have been confused in the contemporary calumnies against poetry and poets; consider how little is, as it appears —or appears, as it is; look to your own motives, and judge not, lest ye be judged.

Poetry, as has been said, differs in this respect from logic, that it is not subject to the control of the active powers of the mind, and that its birth and recurrence have no necessary connexion with the consciousness or will. It is presumptuous to determine that these are the necessary conditions of all mental causation, when mental effects are experienced unsusceptible of being referred to them. The frequent recurrence of the poetical power, it is obvious to suppose, may produce in the mind a habit of order and harmony correlative with its own nature and with its effects upon other minds. But in the intervals of inspiration, and they may be frequent without being durable, a poet becomes a man, and is abandoned to the sudden reflux of the influences under which others habitually live. But as he is more delicately organized than other men, and sensible to pain and pleasure, both his own and that of others, in a degree unknown to them, he will avoid the one and pursue the other with an ardour proportioned to this difference. And he renders himself obnoxious to calumny, when he neglects to observe the circumstances under which these objects of universal pursuit and flight have disguised themselves in one another's garments.

But there is nothing necessarily evil in this error, and thus cruelty, envy, revenge, avarice, and the passions purely evil, have never formed any portion of the popular imputations on the lives of poets.

I have thought it most favourable to the cause of truth to set down these remarks according to the order in which they were suggested to my mind, by a consideration of the subject itself, instead of observing the formality of a polemical reply; but if the view which they contain be just, they will be found to involve a refutation of the arguers against poetry, so far at least as regards the first division of the subject. I can readily conjecture what should have moved the gall of some learned and intelligent writers who quarrel with certain versifiers; I confess myself, like them, unwilling to be stunned by the Theseids of the hoarse Codri of the day. Bavius and Maevius undoubtedly are, as they ever were, insufferable persons. But it belongs to a philosophical critic to distinguish rather than confound.

The first part of these remarks has related to poetry in its elements and principles; and it has been shown, as well as the narrow limits assigned them would permit, that what is called poetry, in a restricted sense, has a common source with all other forms of order and of beauty, according to which the materials of human life are susceptible of being arranged, and which is poetry in a universal sense.

The second part will have for its object an application of these principles to the present state of the cultivation of poetry, and a defence of the attempt to idealize the modern forms of manners and opinions, and compel them into a subordination to the imaginative and creative faculty. For the literature of

England, an energetic development of which has ever preceded or accompanied a great and free development of the national will, has arisen as it were from a new birth. In spite of the low-thoughted envy which would undervalue contemporary merit, our own will be a memorable age in intellectual achievements, and we live among such philosophers and poets as surpass beyond comparison any who have appeared since the last national struggle for civil and religious liberty. The most unfailing herald, companion, and follower of the awakening of a great people to work a beneficial change in opinion or institution, is poetry. At such periods there is an accumulation of the power of communicating and receiving intense and impassioned conceptions respecting man and nature. The persons in whom this power resides may often, as far as regards many portions of their nature, have little apparent correspondence with that spirit of good of which they are the ministers. But even whilst they deny and abjure, they are yet compelled to serve, the power which is seated on the throne of their own soul. It is impossible to read the compositions of the most celebrated writers of the present day without being startled with the electric life which burns within their words. They measure the circumference and sound the depths of human nature with a comprehensive and all-penetrating spirit, and they are themselves perhaps the most sincerely astonished at its manifestations; for it is less their spirit than the spirit of the age. Poets are the hierophants of an unapprehended inspiration; the mirrors of the gigantic shadows which futurity casts upon the present; the words which express what they understand not; the trumpets which sing to battle, and feel not what they inspire; the influence which is moved not, but moves. Poets are the unacknowledged legislators of the world.

1821

# LETTERS

## LETTER I

### To William Godwin.

*Marlow*, 11 *December*, 1817.

I HAVE read and considered all that you say about my general powers, and the particular instance of the poem in which I have attempted to develop them. Nothing can be more satisfactory to me than the interest which your admonitions express. But I think you are mistaken in some points with regard to the peculiar nature of my powers, whatever be their amount. I listened with deference and self-suspicion to your censures of *Laon and Cythna* ; but the productions of mine which you commend hold a very low place in my own esteem, and this reassured me in some degree at least. The poem was produced by a series of thoughts which filled my mind with unbounded and sustained enthusiasm. I felt the precariousness of my life, and I resolved in this book to leave some record of myself. Much of what the volume contains was written with the same feeling, as real, though not so prophetic, as the communications of a dying man. I never presumed, indeed, to consider it anything approaching to faultless ; but, when I considered contemporary productions of the same apparent pretensions, I will own that I was filled with confidence. I felt that in many respects it was a genuine picture of my own mind ; I felt that the sentiments were true, not assumed, and in this have I long believed, that my power consists in sympathy, and that part of imagination which relates to sentiment and contemplation. I am formed, if for anything not in common with the herd of mankind, to apprehend minute and remote distinctions of feeling, whether relative to external nature or the living beings which surround us, and to communicate the conceptions which result from considering either the moral or the material universe as a whole . . . Yet, after all, I cannot but be conscious, in much of what I write, of an absence of that tranquillity which is the attribute and accompaniment of power. This feeling alone would make your most kind and wise admonitions on the subject of the economy of intellectual force valuable to me: and if I live, or if I see any trust in coming years, doubt not but that I shall do something, whatever it may be, which a serious and earnest estimate of my powers will suggest to me, and which will be in every respect accommodated to their utmost limits.

## LETTER II

### To Mr. and Mrs. Gisborne (Leghorn).

*Bagni di Lucca, July 10th*, 1818.

You cannot know, as some friends in England do, to whom my silence is still more inexcusable, that this silence is no proof of forgetfulness or neglect.

I have, in truth, nothing to say, but that I shall be happy to see you again, and renew our delightful walks, until the desire or the duty of seeing new things hurries us away. We have spent a month here in our accustomed solitude, with the exception of one night at the Casino; and the choice society of all ages, which I took care to pack up in a large trunk before we left England, have revisited us here. I am employed just now, having little better to do, in translating into my faint and inefficient periods, the divine eloquence of Plato's *Symposium*; only as an exercise, or, perhaps, to give Mary some idea of the manners and feelings of the Athenians—so different on many subjects from that of any other community that ever existed.

We have almost finished Ariosto—who is entertaining and graceful, and *sometimes* a poet. Forgive me, worshippers of a more equal and tolerant divinity in poetry, if Ariosto pleases me less than you. Where is the gentle seriousness, the delicate sensibility, the calm and sustained energy, without which true greatness cannot be? He is so cruel, too, in his descriptions; his most prized virtues are vices almost without disguise. He constantly vindicates and embellishes revenge in its grossest form; the most deadly superstition that ever infested the world. How different from the tender and solemn enthusiasm of Petrarch—or even the delicate moral sensibility of Tasso, though somewhat obscured by an assumed and artificial style.

We read a good deal here—and we read little in Livorno. We have ridden, Mary and I, once only, to a place called Prato Fiorito, on the top of the mountains: the road, winding through forests, and over torrents, and on the verge of green ravines, affords scenery magnificently fine. I cannot describe it to you, but bid you, though vainly, come and see. I take great delight in watching the changes of the atmosphere here, and the growth of the thunder-showers with which the noon is often overshadowed, and which break and fade away towards evening into flocks of delicate clouds. Our fire-flies are fading away fast; but there is the planet Jupiter, who rises majestically over the rift in the forest-covered mountains to the south, and the pale summer lightning which is spread out every night, at intervals, over the sky. No doubt Providence has contrived these things, that, when the fire-flies go out, the low-flying owl may see her way home.

Remember me kindly to the Machinista.

With the sentiment of impatience until we see you again in the autumn,

I am, yours most sincerely,

P. B. Shelley.

## LETTER III

*Bagni di Lucca, Aug. 16th,* 1818.

My Dear Peacock,—No new event has been added to my life since I wrote last: at least none which might not have taken place as well on the banks of the Thames as on those of the Serchio. I project soon a short excursion, of a week or so, to some of the neighbouring cities; and on the 10th of September we leave this place for Florence, when I shall at least be able to tell you of some things which you cannot see from your windows.

I have finished, by taking advantage of a few days of inspiration—which the *Camoenae* have been lately very backward in conceding—the little poem I began sending to the press in London. Ollier will send you the proofs. Its structure is slight and aëry; its subject ideal. The metre corresponds with the spirit of the poem, and varies with the flow of the feeling. I have translated, and Mary has transcribed, the *Symposium*, as well as my poem; and I am proceeding to employ myself on a discourse, upon the subject of which the *Symposium* treats, considering the subject with reference to the difference of sentiments respecting it, existing between the Greeks and modern nations; a subject to be handled with that delicate caution which either I cannot or I will not practise in other matters, but which here I acknowledge to be necessary. Not that I have any serious thought of publishing either this discourse or the *Symposium*, at least till I return to England, when we may discuss the propriety of it.

*Nightmare Abbey* finished. Well, what is in it? What is it? You are as secret as if the priest of Ceres had dictated its sacred pages. However, I suppose I shall see in time, when my second parcel arrives. My first is yet absent. By what conveyance did you send it?

Pray, are you yet cured of your Nympholepsy? 'Tis a sweet disease: but one as obstinate and dangerous as any—even when the Nymph is a Poliad. Whether such be the case or not, I hope your nympholeptic tale is not abandoned. The subject, if treated with a due spice of Bacchic fury, and interwoven with the manners and feelings of those divine people, who, in their very errors, are the mirrors, as it were, in which all that is delicate and graceful contemplates itself, is perhaps equal to any. What a wonderful passage there is in *Phaedrus*—the beginning, I think, of one of the speeches of Socrates—in praise of poetic madness, and in definition of what poetry is, and how a man becomes a poet. Every man who lives in this age and desires to write poetry, ought, as a preservative against the false and narrow systems of criticism which every poetical empiric vents, to impress himself with this sentence, if he would be numbered among those to whom may apply this proud, though sublime, expression of Tasso: *Non c'è in mondo chi merita nome di creatore, che Dio ed il Poeta.*

The weather has been brilliantly fine; and now, among these mountains, the autumnal air is becoming less hot, especially in the mornings and evenings. The chestnut woods are now inexpressibly beautiful, for the chestnuts have become large, and add a new richness to the full foliage. We see here Jupiter in the east; and Venus, I believe, as the evening star, directly after sunset.

More and better in my next. Mary and Claire desire their kind remembrances. Most faithfully your friend,

P. B. SHELLEY.

## LETTER IV

### TO T. L. P., ESQ.

*Bologna; Monday, Nov. 9th*, 1818.

MY DEAR P.,—I have seen a quantity of things here —churches, palaces, statues, fountains, and pictures; and my brain is at this moment like a portfolio of an architect, or a print-shop, or a commonplace-book. I will try to recollect something of what I have seen; for, indeed, it requires, if it will obey, an act of volition. First, we went to the cathedral, which contains nothing remarkable, except a kind of shrine, or rather a marble canopy, loaded with sculptures, and supported on four marble columns. We went then to a palace—I am sure I forget the name of it—where we saw a large gallery of pictures. Of course, in a picture gallery you see three hundred pictures you forget, for one you remember. I remember, however, an interesting picture by Guido, of the Rape of Proserpine, in which Proserpine casts back her languid and half-unwilling eyes, as it were, to the flowers she had left ungathered in the fields of Enna. There was an exquisitely executed piece of Correggio, about four saints, one of whom seemed to have a pet dragon in a leash. I was told that it was the devil who was bound in that style—but who can make anything of four saints ? For what can they be supposed to be about ? There was one painting, indeed, by this master, Christ beatified, inexpressibly fine. It is a half figure, seated on a mass of clouds, tinged with an ethereal, rose-like lustre ; the arms are expanded ; the whole frame seems dilated with expression ; the countenance is heavy, as it were, with the weight of the rapture of the spirit ; the lips parted, but scarcely parted, with the breath of intense but regulated passion ; the eyes are calm and benignant ; the whole features harmonize in majesty and sweetness. The hair is parted on the forehead, and falls in heavy locks on each side. It is motionless, but seems as if the faintest breath would move it. The colouring, I suppose, must be very good, if I could remark and understand it. The sky is of a pale aerial orange, like the tints of latest sunset ; it does not seem painted around and beyond the figure, but everything seems to have absorbed, and to have been penetrated by its hues. I do not think we saw any other of Correggio, but this specimen gives me a very exalted idea of his powers.

We went to see heaven knows how many more palaces—Ranuzzi, Marriscalchi, Aldobrandi. If you want Italian names for any purpose, here they are ; I should be glad of them if I was writing a novel. I saw many more of Guido. One, a Samson drinking water out of an ass's jaw-bone, in the midst of the slaughtered Philistines. Why he is supposed to do this, God, who gave him this jaw-bone, alone knows—but certain it is, that the painting is a very fine one.

The figure of Samson stands in strong relief in the fore-ground, coloured, as it were, in the hues of human life, and full of strength and elegance. Round him lie the Philistines in all the attitudes of death. One prone, with the slight convulsion of pain just passing from his forehead, whilst on his lips and chin death lies as heavy as sleep. Another leaning on his arm, with his hand, white and motionless, hanging out beyond. In the distance, more dead bodies; and still further beyond, the blue sea and the blue mountains, and one white and tranquil sail.

There is a Murder of the Innocents, also, by Guido, finely coloured, with much fine expression—but the subject is very horrible, and it seemed deficient in strength,—at least, you require the highest ideal energy, the most poetical and exalted conception of the subject, to reconcile you to such a contemplation. There was a Jesus Christ crucified, by the same, very fine. One gets tired, indeed, whatever may be the conception and execution of it, of seeing that monotonous and agonized form for ever exhibited in one prescriptive attitude of torture. But the Magdalen, clinging to the cross with the look of passive and gentle despair beaming from beneath her bright flaxen hair, and the figure of St. John, with his looks uplifted in passionate compassion; his hands clasped, and his fingers twisting themselves together, as it were, with involuntary anguish; his feet almost writhing up from the ground with the same sympathy; and the whole of this arrayed in colours of a diviner nature, yet most like nature's self. Of the contemplation of this one would never weary.

There was a 'Fortune', too, of Guido; a piece of mere beauty. There was the figure of Fortune on a globe, eagerly proceeding onwards, and Love was trying to catch her back by the hair, and her face was half turned towards him; her long chestnut hair was floating in the stream of the wind, and threw its shadow over her fair forehead. Her hazel eyes were fixed on her pursuer, with a meaning look of playfulness, and a light smile was hovering on her lips. The colours which arrayed her delicate limbs were ethereal and warm.

But, perhaps, the most interesting of all the pictures of Guido which I saw was a Madonna Lattante. She is leaning over her child, and the maternal feelings with which she is pervaded are shadowed forth on her soft and gentle countenance, and in her simple and affectionate gestures—there is what an unfeeling observer would call a dullness in the expression of her face; her eyes are almost closed; her lip depressed; there is a serious, and even a heavy relaxation, as it were, of all the muscles which are called into action by ordinary emotions: but it is only as if the spirit of love, almost insupportable from its intensity, were brooding over and weighing down the soul, or whatever it is, without which the material frame is inanimate and inexpressive.

There is another painter here, called Franceschini, a Bolognese, who, though certainly very inferior to Guido, is yet a person of excellent powers. One entire church, that of Santa Catarina, is covered by his works. I do not know whether any of his pictures have ever been seen in England. His colouring is less warm than that of Guido, but nothing can be more clear and delicate; it is as if he could have dipped his pencil in the hues of some serenest and star-shining twilight. His forms have the same delicacy and aerial loveliness; their eyes are all bright with innocence and love; their lips scarce divided by some gentle and sweet emotion. His winged children are the loveliest ideal beings ever created by the human mind. These are generally, whether in the capacity of Cherubim or Cupid, accessories to the rest of the picture ; and the underplot of their lovely and infantine play is something almost pathetic, from the excess of its unpretending beauty. One of the best of his pieces is an Annunciation of the Virgin :—the Angel is beaming in beauty ; the Virgin, soft, retiring, and simple.

We saw, besides, one picture of Raphael—St. Cecilia : this is in another and higher style ; you forget that it is a picture as you look at it ; and yet it is most unlike any of those things which we call reality. It is of the inspired and ideal kind, and seems to have been conceived and executed in a similar state of feeling to that which produced among the ancients those perfect specimens of poetry and sculpture which are the baffling models of succeeding generations. There is a unity and perfection in it of an incommunicable kind. The central figure, St. Cecilia, seems rapt in such inspiration as produced her image in the painter's mind ; her deep, dark, eloquent eyes lifted up ; her chestnut hair flung back from her forehead—she holds an organ in her hands—her countenance, as it were, calmed by the depth of its passion and rapture, and penetrated throughout with the warm and radiant light of life. She is listening to the music of heaven, and, as I imagine, has just ceased to sing, for the four figures that surround her evidently point, by their attitudes, towards her ; particularly St. John, who, with a tender yet impassioned gesture, bends his countenance towards her, languid with the depth of his emotion. At her feet lie various instruments of music, broken and unstrung. Of the colouring I do not speak ; it eclipses Nature, yet it has all her truth and softness.

We saw some pictures of Domenichino, Caracci, Albano, Guercino, Elizabetta Sirani. The two former, remember, I do not pretend to taste — I cannot admire. Of the latter there are some beautiful Madonnas. There are several of Guercino, which they said were very fine. I dare say they were, for the strength and complication of his figures made my head turn round. One, indeed, was certainly powerful. It was the representation of the founder of the Carthusians exercising his austerities in the desert, with a youth as his attendant, kneeling beside him at an altar; on another altar stood a skull and a crucifix; and around were the rocks and the trees of the wilderness. I never saw such a figure as this fellow. His face was wrinkled like a dried snake's skin, and drawn in long hard lines: his very hands were wrinkled. He looked like an animated mummy. He was clothed in a loose dress of death-coloured flannel, such as you might fancy a shroud might be, after it had wrapped a corpse a month or two. It had a yellow, putrefied, ghastly hue, which it cast on all the objects around, so that the hands and face of the Carthusian and his companion were jaundiced by this sepulchral glimmer. Why write books against religion, when we may hang up such pictures? But the world either will not or cannot see. The gloomy effect of this was softened, and, at the same time, its sublimity diminished, by the figure of the Virgin and Child in the sky, looking down with admiration on the monk, and a beautiful flying figure of an angel.

Enough of pictures. I saw the place where Guido and his mistress, Elizabetta Sirani, were buried. This lady was poisoned at the age of twenty-six, by another lover, a rejected one of course. Our guide said she was very ugly, and that we might see her portrait to-morrow.

Well, good-night, for the present. 'To-morrow to fresh fields and pastures new.'

*Nov.* 16.

To-day we first went to see those divine pictures of Raffael and Guido again, and then rode up the mountains, behind this city, to visit a chapel dedicated to the Madonna. It made me melancholy to see that they had been varnishing and restoring some of these pictures, and that even some had been pierced by the French bayonets. These are symptoms of the mortality of man, and, perhaps, few of his works are more evanescent than paintings. Sculpture retains its freshness for twenty centuries—the Apollo and the Venus are as they were. But books are perhaps the only productions of man coeval with the human race. Sophocles and Shakespeare can be produced and re-produced for ever. But how evanescent are paintings! and must necessarily be. Those of Zeuxis and Apelles are no more; and perhaps they bore the same relation to Homer and Aeschylus, that those of Guido and Raffael bear to Dante and Petrarch. There is one refuge from the despondency of this contemplation. The material part, indeed, of their works must perish, but they survive in the mind of man, and the remembrances connected with them are transmitted from generation to generation. The poet embodies them in his creations; the systems of philosophers are modelled to gentleness by their contemplation; opinion, that legislator, is infected with their influence; men become better and wiser; and the unseen seeds are perhaps thus sown, which shall produce a plant more excellent even than that from which they fell. But all this might as well be said or thought at Marlow as Bologna.

The chapel of the Madonna is a very pretty Corinthian building—very beautiful indeed. It commands a fine view of these fertile plains, the many-folded Apennines, and the city. I have just returned from a moonlight walk through Bologna. It is a city of colonnades, and the effect of moonlight is strikingly picturesque. There are two towers here—one 400 feet high—ugly things, built of brick, which lean both different ways; and with the delusion of moonlight shadows, you might almost fancy that the city is rocked by an earthquake. They say they were built so on purpose; but I observe in all the plain of Lombardy the church towers lean.

Adieu.—God grant you patience to read this long letter, and courage to support the expectation of the next. Pray part them from the *Cobbetts* on your breakfast table—they may fight it out in your mind.

Yours ever, most sincerely,

P. B. S.

## LETTER V

### To Thomas Love Peacock.

*Naples, February 25th*, 1819.

My dear Peacock,—I am much interested to hear your progress in the object of your removal to London, especially as I hear from Horace Smith of the advantages attending it. There is no person in the world who would more sincerely rejoice in any good fortune that might befall you than I should.

We are on the point of quitting Naples for Rome. The scenery which surrounds this city is more delightful than any within the immediate reach of civilized man. I don't think I have mentioned to you the Lago d'Agnano and the Caccia d'Ischieri, and I have since seen what obscures those lovely forms in my memory. They are both the craters of extinguished volcanoes, and Nature has thrown forth forests of oak and ilex, and spread mossy lawns and clear lakes over the dead or sleeping fire. The first is a scene of a wider and milder character, with soft sloping, wooded hills, and grassy declivities declining to the lake, and cultivated plains of vines woven upon poplar-trees, bounded by the theatre of hills. Innumerable wild water-birds, quite tame, inhabit this place. The other is a royal chace, is surrounded by steep and lofty hills, and only accessible through a wide gate of massy oak, from the vestibule of which the spectacle of precipitous hills, hemming in a narrow and circular vale, is suddenly disclosed. The hills are covered with thick woods of ilex, myrtle, and laurustinus ; the polished leaves of the ilex, as they wave in their multitudes under the partial blasts which rush through the chasms of the vale, glitter above the dark masses of foliage below, like the white foam of waves upon a deep blue sea. The plain so surrounded is at most three miles in circumference. It is occupied partly by a lake, with bold shores wooded by evergreens, and interrupted by a sylvan promontory of the wild forest, whose mossy boughs overhang its expanse, of a silent and purple darkness, like an Italian midnight ; and partly by the forest itself, of all gigantic trees, but the oak especially, whose jagged boughs, now leafless, are hoary with thick lichens, and loaded with the massy and deep foliage of the ivy. The effect of the dark eminences that surround this plain, seen through the boughs, is of an enchanting solemnity. (There we saw in one instance wild boars and a deer, and in another—a spectacle little suited to the antique and Latonian nature of the place—King Ferdinand in a winter enclosure, watching to shoot wild boars.) The underwood was principally evergreen, all lovely kinds of fern and furze; the cytisus, a delicate kind of furze with a pretty yellow blossom, the myrtle, and the myrica. The willow-trees had just begun to put forth their green and golden buds, and gleamed like points of lambent fire among the wintry forest. The Grotto del Cane, too, we saw, because other people see it; but would not allow the dog to be exhibited in torture for our curiosity. The poor little animals stood moving their tails in a slow and dismal manner, as if perfectly resigned to their condition—a cur-like emblem of voluntary servitude. The effect of the vapour, which extinguishes a torch, is to cause suffocation at last, through a process which makes the lungs feel as if they were torn by sharp points within. So a surgeon told us, who tried the experiment on himself.

There was a Greek city, sixty miles to the south of Naples, called Posidonia, now Pesto, where there still subsist three temples of Etruscan architecture, still perfect. From this city we have just returned. The weather was most unfavourable for our expedition. After two months of cloudless serenity, it began raining cats and dogs. The first night we slept at Salerno, a large city situate in the recess of a deep bay; surrounded with stupendous mountains of the same name. A few miles from Torre del Greco we entered on the pass of the mountains, which is a line dividing the isthmus of those enormous piles of rock which compose the southern boundary of the Bay of Naples, and the northern one of that of Salerno. On one side is a lofty conical hill, crowned with the turrets of a ruined castle, and cut into platforms for cultivation; at least every ravine and glen, whose precipitous sides admitted of other vegetation but that of the rock-rooted ilex; on the other the ethereal snowy crags of an immense mountain, whose terrible lineaments were at intervals concealed or disclosed by volumes of dense clouds rolling under the tempest. Half a mile from this spot, between orange and lemon groves of a lovely village, suspended as it were on an amphitheatral precipice, whose golden globes contrasted with the white walls and dark green leaves which they almost outnumbered, shone the sea. A burst of the declining sunlight illumined it. The road led along the brink of the precipice, towards Salerno. Nothing could be more glorious than the scene. The immense mountains covered with the rare and divine vegetation of this climate, with many-folding vales, and deep dark recesses, which the fancy scarcely could penetrate, descended from their snowy summits precipitously to the sea. Before us was Salerno, built into a declining plain, between the mountains and the sea. Beyond, the other shore of sky-cleaving mountains, then dim with the mist of tempest. Underneath, from the base of the precipice where the road conducted, rocky promontories jutted into the sea, covered with olive and ilex woods, or with the ruined battlements of some Norman or Saracenic fortress. We slept at Salerno, and the next morning, before daybreak, proceeded to Posidonia. The night had been tempestuous, and our way lay by the sea sand. It was utterly dark, except when the long line of wave burst, with a sound like thunder, beneath the starless sky, and cast up a kind of mist of cold white lustre. When morning came, we found ourselves travelling in a wide desert plain, perpetually interrupted by wild irregular glens, and bounded on all sides by the Apennines and the sea. Sometimes it was covered with forest, sometimes dotted with underwood, or mere tufts of fern and furze, and the wintry dry tendrils of creeping plants. I have never, but in the Alps, seen an amphitheatre of mountains so magnificent. After travelling fifteen miles, we came to a river, the bridge of which had been broken, and which was so swollen that the ferry would not take the carriage across. We had, therefore, to walk seven miles of a muddy road, which led to the ancient city across the desolate Maremma. The air was scented with the sweet smell of violets of an extraordinary size and beauty. At length we saw the sublime and massive colonnades, skirting the horizon of the wilderness. We entered by the ancient gate, which is now no more than a chasm in the rock-like wall. Deeply sunk in the ground beside it were the ruins of a sepulchre, which the ancients were in the custom of building beside the public way. The first temple, which is the smallest, consists of an outer range of columns, quite perfect, and supporting a perfect architrave and two shattered frontispieces. The proportions are extremely massy, and the architecture entirely unornamented and simple. These columns do not seem more than forty feet high, but the perfect proportions diminish the apprehension of their magnitude; it seems as if inequality and irregularity of form were requisite to force on us the relative idea of greatness. The scene from between the columns of the temple consists on one side of the sea, to which the gentle hill on which it is built slopes, and on the other, of the grand amphitheatre of the loftiest Apennines, dark purple mountains, crowned with snow, and intersected there by long bars of hard and leaden-coloured cloud. The effect of the jagged outline of mountains, through groups of enormous columns on one side, and on the other the level horizon of the sea, is inexpressibly grand. The second temple is much larger, and also more perfect. Beside the outer range of columns, it contains an interior range of column above column, and the ruins of a wall which was the screen of the penetralia. With little diversity of ornament, the order of architecture is similar to that of the first temple. The columns in all are fluted, and built of a porous volcanic stone, which time has dyed with a rich and yellow colour. The columns are one-third larger, and like that of the first, diminish from the base to the capital, so that, but for the chastening effect of their admirable proportions, their magnitude would, from the delusion of perspective, seem greater, not less, than it is; though perhaps we ought to say, not that this symmetry diminishes your apprehension of their magnitude, but that it overpowers the idea of relative greatness, by establishing within itself a system of relations destructive of your idea of its relation with other objects, on which our ideas of size depend. The third temple is what they call a Basilica; three columns alone remain of the interior range; the exterior is perfect, but that the cornice and frieze in many places have fallen. This temple covers more ground than either of the others, but its columns are of an intermediate magnitude between those of the second and the first.

We only contemplated these sublime monuments for two hours, and of course could only bring away so imperfect a conception of them as is the shadow of some half-remembered dream.

The royal collection of paintings in this city is sufficiently miserable. Perhaps the most remarkable is the original studio by Michael Angelo, of the 'Day of Judgement', which is painted in fresco on the Sixtine chapel of the Vatican. It is there so defaced as to be wholly indistinguishable. I cannot but think the genius of this artist highly overrated. He has not only no temperance, no modesty, no feeling for the just boundaries of art (and in these respects an admirable genius may err), but he has no sense of beauty, and to want this is to want the sense of the creative power of mind. What is terror without a contrast with, and a connexion with, loveliness? How well Dante understood this secret—Dante, with whom this artist has been so presumptuously compared! What a thing his 'Moses' is; how distorted from all that is natural and majestic, only less monstrous and detestable than its historical prototype. In the picture to which I allude, God is leaning out of heaven, as it were eagerly enjoying the final scene of the infernal tragedy he set the Universe to act. The Holy Ghost, in the shape of a dove, is under Him. Under the Holy Ghost stands Jesus Christ, in an attitude of haranguing the assembly. This figure, which his subject, or rather the view which it became him to take of it, ought to have modelled of a calm, severe, awe-inspiring majesty, terrible yet lovely, is in the attitude of commonplace resentment. On one side of this figure are the elect; on the other, the host of heaven; they ought to have been what the Christians call *glorified bodies*, floating onward and radiant with that everlasting light (I speak in the spirit of their faith), which had consumed their mortal veil. They are in fact very ordinary people. Below is the ideal purgatory, I imagine, in mid-air, in the shapes of spirits, some of whom demons are dragging down, others falling as it were by their own weight, others half suspended in that Mahomet-coffin kind of attitude which most moderate Christians, I believe, expect to assume. Every step towards hell approximates to the region of the artist's exclusive power. There is great imagination in many of the situations of these unfortunate spirits. But hell and death are his real sphere. The bottom of the picture is divided by a lofty rock, in which there is a cavern whose entrance is thronged by devils, some coming in with spirits, some going out for prey. The blood-red light of the fiery abyss glows through their dark forms. On one side are the devils in all hideous forms, struggling with the damned, who have received their sentence at the redeemer's throne, and chained in all forms of agony by knotted serpents, and writhing on the crags in every variety of torture. On the other, are the dead coming out of their graves—horrible forms. Such is the famous 'Day of Judgement' of Michael Angelo; a kind of *Titus Andronicus* in painting, but the author surely no Shakespeare. The other paintings are one or two of Raphael or his pupils, very sweet and lovely. A 'Danaë', of Titian, a picture, the softest and most voluptuous form, with languid and uplifted eyes, and warm yet passive limbs. A 'Maddelena', by Guido, with dark brown hair, and dark brown eyes, and an earnest, soft, melancholy look. And some excellent pictures, in point of execution, by Annibal Carracci. None others worth a second look. Of the gallery of statues I cannot speak. They require a volume, not a letter. Still less what can I do at Rome?

I have just seen the *Quarterly* for September (not from my own box). I suppose there is no chance now of your organizing a review. This is a great pity. The *Quarterly* is undoubtedly conducted with talent, great talent, and affords a dreadful preponderance against the cause of improvement. If a band of staunch reformers, resolute yet skilful infidels, were united in so close and constant a league as that in which interest and fanaticism have bound the members of that literary coalition!

Adieu. Address your next letter to Rome, whence you shall hear from me soon again. Mary and Clara unite with me in the very kindest remembrances.

Most faithfully yours,

P. B. S.

A doctor here has been messing me, and I believe has done me an important benefit. One of his pretty schemes has been putting caustic on my side. You may guess how much quiet I have had since it was laid on. . . . . .

## LETTER VI

To T. L. P., Esq.

*Rome, March 23rd*, 1819.

My dear P.,—I wrote to you the day before our departure from Naples. We came by slow journeys, with our own horses, to Rome, resting one day at Mola di Gaeta, at the inn called Villa di Cicerone, from being built on the ruins of his Villa, whose immense substructions overhang the sea, and are scattered among the orange groves. Nothing can be lovelier than the scene from the terraces of the inn. On one side precipitous mountains, whose bases slope into an inclined plane of olive and orange copses—the latter forming, as it were, an emerald sky of leaves, starred with innumerable globes of their ripening fruit, whose rich splendour contrasted with the deep green foliage ; on the other the sea—bounded on one side by the antique town of Gaeta, and the other by what appears to be an island, the promontory of Circe. From Gaeta to Terracina the whole scenery is of the most sublime character. At Terracina, precipitous conical crags of immense height shoot into the sky and overhang the sea. At Albano, we arrived again in sight of Rome. Arches after arches in unending lines stretching across the uninhabited wilderness, the blue defined line of the mountains seen between them ; masses of nameless ruin standing like rocks out of the plain ; and the plain itself, with its billowy and unequal surface, announced the neighbourhood of Rome. And what shall I say to you of Rome ? If I speak of the inanimate ruins, the rude stones piled upon stones, which are the sepulchres of the fame of those who once arrayed them with the beauty which has faded, will you believe me insensible to the vital, the almost breathing creations of genius yet subsisting in their perfection ? What has become, you will ask, of the Apollo, the Gladiator, the Venus of the Capitol ? What of the Apollo di Belvedere, the Laocoön ? What of Raffael and Guido ? These things are best spoken of when the mind has drunk in the spirit of their forms ; and little indeed can I, who must devote no more than a few months to the contemplation of them, hope to know or feel of their profound beauty.

I think I told you of the Coliseum, and its impressions on me on my first visit to this city. The next most considerable relic of antiquity, considered as a ruin, is the Thermæ of Caracalla. These consist of six enormous chambers, above 200 feet in height, and each enclosing a vast space like that of a field. There are, in addition, a number of towers and labyrinthine recesses, hidden and woven over by the wild growth of weeds and ivy. Never was any desolation more sublime and lovely. The perpendicular wall of ruin is cloven into steep ravines filled up with flowering shrubs, whose thick twisted roots are knotted in the rifts of the stones. At every step the aerial pinnacles of shattered stone group into new combinations of effect, and tower above the lofty yet level walls, as the distant mountains change their aspect to one travelling rapidly along the plain. The perpendicular walls resemble nothing more than that cliff of Bisham wood, that is overgrown with wood, and yet is stony and precipitous—you know the one I mean; not the chalk-pit, but the spot that has the pretty copse of fir-trees and privet-bushes at its base, and where H—— and I scrambled up, and you, to my infinite discontent, would go home. These walls surround green and level spaces of lawn, on which some elms have grown, and which are interspersed towards their skirts by masses of the fallen ruin, overtwined with the broad leaves of the creeping weeds. The blue sky canopies it, and is as the everlasting roof of these enormous halls.

But the most interesting effect remains. In one of the buttresses, that supports an immense and lofty arch, 'which bridges the very winds of heaven,' are the crumbling remains of an antique winding staircase, whose sides are open in many places to the precipice. This you ascend, and arrive on the summit of these piles. There grow on every side thick entangled wildernesses of myrtle, and the myrletus, and bay, and the flowering laurestinus, whose white blossoms are just developed, the white fig, and a thousand nameless plants sown by the wandering winds. These woods are intersected on every side by paths, like sheep-tracks through the copse-wood of steep mountains, which wind to every part of the immense labyrinth. From the midst rise those pinnacles and masses, themselves like mountains, which have been seen from below. In one place you wind along a narrow strip of weed-grown ruin: on one side is the immensity of earth and sky, on the other a narrow chasm, which is bounded by an arch of enormous size, fringed by the many-coloured foliage and blossoms, and supporting a lofty and irregular pyramid, overgrown like itself with the all-prevailing vegetation. Around rise other crags and other peaks, all arrayed, and the deformity of their vast desolation softened down, by the undecaying investiture of nature. Come to Rome. It is a scene by which expression is overpowered ; which words cannot convey. Still further, winding up one half of the shattered pyramids, by the path through the blooming copsewood, you come to a little mossy lawn, surrounded by the wild shrubs ; it is overgrown with anemonies, wall-flowers, and violets, whose stalks pierce the starry moss, and with radiant blue flowers, whose names I know not, and which scatter through the air the divinest odour, which, as you recline under the shade of the ruin, produces sensations of voluptuous faintness, like the combinations of sweet music. The paths still wind on, threading the perplexed windings, other labyrinths, other lawns, and deep dells of wood, and lofty rocks, and terrific chasms. When I tell you that these ruins cover several acres, and that the paths above penetrate at least half their extent, your imagination will fill up all that I am unable to express of this astonishing scene.

I speak of these things not in the order in which I visited them, but in that of the impression which they made on me, or perhaps chance directs. The ruins of the ancient Forum are so far fortunate that they have not been walled up in the modern city. They stand in an open, lonesome place, bounded on one side by the modern city, and the other by the Palatine Mount, covered with shapeless masses of ruin. The tourists tell you all about these things, and I am afraid of stumbling on their language when I enumerate what is so well known. There remain eight granite columns of the Ionic order, with their entablature, of the temple of Concord, founded by Camillus. I fear that the immense expense demanded by these columns forbids us to hope that they are the remains of any edifice dedicated by that most perfect and virtuous of men. It is supposed to have been repaired under the Eastern Emperors; alas, what a contrast of recollections! Near them stand those Corinthian fluted columns, which supported the angle of a temple; the architrave and entablature are worked with delicate sculpture. Beyond, to the south, is another solitary column; and still more distant, three more, supporting the wreck of an entablature. Descending from the Capitol to the Forum, is the triumphal arch of Septimius Severus, less perfect than that of Constantine, though from its proportions and magnitude a most impressive monument. That of Constantine, or rather of Titus, (for the relief and sculpture, and even the colossal images of Dacian captives, were torn by a decree of the senate from an arch dedicated to the latter, to adorn that of this stupid and wicked monster, Constantine, one of whose chief merits consists in establishing a religion, the destroyer of those arts which would have rendered so base a spoliation unnecessary) is the most perfect. It is an admirable work of art. It is built of the finest marble, and the outline of the reliefs is in many parts as perfect as if just finished. Four Corinthian fluted columns support, on each side, a bold entablature, whose bases are loaded with reliefs of captives in every attitude of humiliation and slavery. The compartments above express, in bolder relief, the enjoyment of success; the conqueror on his throne, or in his chariot, or nodding over the crushed multitudes, who writhe under his horses' hoofs, as those below express the torture and abjectness of defeat. There are three arches, whose roofs are panelled with fretwork, and their sides adorned with similar reliefs. The keystone of these arches is supported each by two winged figures of Victory, whose hair floats on the wind of their own speed, and whose arms are outstretched, bearing trophies, as if impatient to meet. They look, as it were, borne from the subject extremities of the earth, on the breath which is the exhalation of that battle and desolation, which it is their mission to commemorate. Never were monuments so completely fitted to the purpose for which they were designed, of expressing that mixture of energy and error which is called a triumph.

I walk forth in the purple and golden light of an Italian evening, and return by star or moon light, through this scene. The elms are just budding, and the warm spring winds bring unknown odours, all sweet, from the country. I see the radiant Orion through the mighty columns of the temple of Concord, and the mellow fading light softens down the modern buildings of the Capitol, the only ones that interfere with the sublime desolation of the scene. On the steps of the Capitol itself, stand two colossal statues of Castor and Pollux, each with his horse, finely executed, though far inferior to those of Monte Cavallo, the cast of one of which you know we saw together in London. This walk is close to our lodging, and this is my evening walk.

What shall I say of the modern city ? Rome is yet the capital of the world. It is a city of palaces and temples, more glorious than those which any other city contains, and of ruins more glorious than they. Seen from any of the eminences that surround it, it exhibits domes beyond domes, and palaces, and colonnades interminably, even to the horizon ; interspersed with patches of desert, and mighty ruins which stand girt by their own desolation, in the midst of the fanes of living religions and the habitations of living men, in sublime loneliness. St. Peter's is, as you have heard, the loftiest building in Europe. Externally it is inferior in architectural beauty to St. Paul's, though not wholly devoid of it; internally it exhibits littleness on a large scale, and is in every respect opposed to antique taste. You know my propensity to admire; and I tried to persuade myself out of this opinion—in vain; the more I see of the interior of St. Peter's, the less impression as a whole does it produce on me. I cannot even think it lofty, though its dome is considerably higher than any hill within fifty miles of London; and when one reflects, it is an astonishing monument of the daring energy of man. Its colonnade is wonderfully fine, and there are two fountains, which rise in spire-like columns of water to an immense height in the sky, and falling on the porphyry vases from which they spring, fill the whole air with a radiant mist, which at noon is thronged with innumerable rainbows. In the midst stands an obelisk. In front is the palace-like façade of St. Peter's, certainly magnificent; and there is produced, on the whole, an architectural combination unequalled in the world. But the dome of the temple is concealed, except at a very great distance, by the façade and the inferior part of the building, and that diabolical contrivance they call an attic.

The effect of the Pantheon is totally the reverse of that of St. Peter's. Though not a fourth part of the size, it is, as it were, the visible image of the universe; in the perfection of its proportions, as when you regard the unmeasured dome of heaven, the idea of magnitude is swallowed up and lost. It is open to the sky, and its wide dome is lighted by the ever-changing illumination of the air. The clouds of noon fly over it, and at night the keen stars are seen through the azure darkness, hanging immovably, or driving after the driving moon among the clouds. We visited it by moonlight; it is supported by sixteen columns, fluted and Corinthian, of a certain rare and beautiful yellow marble, exquisitely polished, called here *giallo antico*. Above these are the niches for the statues of the twelve gods. This is the only defect of this sublime temple; there ought to have been no interval between the commencement of the dome and the cornice, supported by the columns. Thus there would have been no diversion from the magnificent simplicity of its form. This improvement is alone wanting to have completed the unity of the idea.

The fountains of Rome are, in themselves, magnificent combinations of art, such as alone it were worth coming to see. That in the Piazza Navona, a large square, is composed of enormous fragments of rock, piled on each other, and penetrated as by caverns. This mass supports an Egyptian obelisk of immense height. On the four corners of the rock recline, in different attitudes, colossal figures representing the four divisions of the globe. The water bursts from the crevices beneath them. They are sculptured with great spirit; one impatiently tearing a veil from his eyes; another with his hands stretched upwards. The Fontana di Trevi is the most celebrated, and is rather a waterfall than a fountain; gushing out from masses of rock, with a gigantic figure of Neptune; and below are two river gods, checking two winged horses, struggling up from among the rocks and waters. The whole is not ill conceived nor executed; but you know not how delicate the imagination becomes by dieting with antiquity day after day! The only things that sustain the comparison are Raffael, Guido, and Salvator Rosa.

The fountain on the Quirinal, or rather the group formed by the statues, obelisk and the fountain, is, however, the most admirable of all. From the Piazza Quirinale, or rather Monte Cavallo, you see the boundless ocean of domes, spires, and columns, which is the City, Rome. On a pedestal of white marble rises an obelisk of red granite, piercing the blue sky. Before it is a vast basin of porphyry, in the midst of which rises a column of the purest water, which collects into itself all the overhanging colours of the sky, and breaks them into a thousand prismatic hues and graduated shadows—they fall together with its dashing water-drops into the outer basin. The elevated situation of this fountain produces, I imagine, this effect of colour. On each side, on an elevated pedestal, stand the statues of Castor and Pollux, each in the act of taming his horse; which are said, but I believe wholly without authority, to be the work of Phidias and Praxiteles. These figures combine the irresistible energy with the sublime and perfect loveliness supposed to have belonged to their divine nature. The reins no longer exist, but the position of their hands and the sustained and calm command of their regard, seem to require no mechanical aid to enforce obedience. The countenances at so great a height are scarcely visible, and I have a better idea of that of which we saw a cast together in London, than of the other. But the sublime and living majesty of their limbs and mien, the nervous and fiery animation of the horses they restrain, seen in the blue sky of Italy, and overlooking the city of Rome, surrounded by the light and the music of that crystalline fountain, no cast can communicate.

These figures were found at the Baths of Constantine; but, of course, are of remote antiquity. I do not acquiesce, however, in the practice of attributing to Phidias, or Praxiteles, or Scopas, or some great master, any admirable work that may be found. We find little of what remained, and perhaps the works of these were such as greatly surpassed all that we conceive of most perfect and admirable in what little has escaped the *deluge.* If I am too jealous of the honour of the Greeks, our masters and creators, the gods whom we should worship,—pardon me.

I have said what I feel without entering into any critical discussions of the *ruins* of Rome, and the mere outside of this inexhaustible mine of thought and feeling. Hobhouse, Eustace, and Forsyth, will tell all the show-knowledge about it,—'the common stuff of the earth'. By-the-by, Forsyth is worth reading, as I judge from a chapter or two I have seen. I cannot get the book here.

I ought to have observed that the central arch of the triumphal Arch of Titus yet subsists, more perfect in its proportions, they say, than any of a later date. This I did not remark. The figures of Victory, with unfolded wings, and each spurning back a globe with outstretched feet, are, perhaps, more beautiful than those on either of the others. Their lips are parted: a delicate mode of indicating the fervour of their desire to arrive at the destined resting-place, and to express the eager respiration of their speed. Indeed, so essential to beauty were the forms expressive of the exercise of the imagination and the affections considered by *Greek* artists, that no ideal figure of antiquity, not destined to some representation directly exclusive of such a character, is to be found with closed lips. Within this arch are two panelled alto-relievos, one representing a train of people bearing in procession the instruments of Jewish worship, among which is the holy candlestick with seven branches ; on the other, Titus standing on a quadriga, with a winged Victory. The grouping of the horses, and the beauty, correctness, and energy of their delineation, is remarkable, though they are much destroyed.

## LETTER VII

### To T. L. P., Esq.

*Livorno, July*, 1819.

My dear P.—We still remain, and shall remain nearly two months longer, at Livorno. Our house is a melancholy one,[1] and only cheered by letters from England. I got your note, in which you speak of three letters having been sent to Naples, which I have written for. I have heard also from H——, who confirms the news of your success, an intelligence most grateful to me.

The object of the present letter is to ask a favour of you. I have written a tragedy, on the subject of a story well known in Italy, and, in my conception, eminently dramatic. I have taken some pains to make my play fit for representation, and those who have already seen it judge favourably. It is written without any of the peculiar feelings and opinions which characterize my other compositions ; I having attended simply to the impartial development of such characters, as it is probable the persons represented really were, together with the greatest degree of popular effect to be produced by such a development. I send you a translation of the Italian manuscript on which my play is founded, the chief subject

[1] We had lost our eldest, and, at that time, only child, the preceding month at Rome.

of which I have touched very delicately; for my principal doubt, as to whether it would succeed as an acting play, hangs entirely on the question, as to whether such a thing as incest in this shape, however treated, would be admitted on the stage. I think, however, it will form no objection: considering, first, that the facts are matter of history; and, secondly, the peculiar delicacy with which I have treated it.

I am exceedingly interested in the question of whether this attempt of mine will succeed or no. I am strongly inclined to the affirmative at present, founding my hopes on this, that, as a composition, it is certainly not inferior to any of the modern plays that have been acted, with the exception of *Remorse*; that the interest of its plot is incredibly greater and more real; and that there is nothing beyond what the multitude are contented to believe that they can understand, either in imagery, opinion, or sentiment. I wish to preserve a complete incognito, and can trust to you, that whatever else you do, you will at least favour me on this point. Indeed this is essential, deeply essential to its success. After it had been acted, and successfully (could I hope such a thing), I would own it if I pleased, and use the celebrity it might acquire to my own purposes.

What I want you to do is, to procure for me its presentation at Covent Garden. The principal character, Beatrice, is precisely fitted for Miss O'Neil, and it might even seem written for her, (God forbid that I should ever see her play it—it would tear my nerves to pieces,) and, in all respects, it is fitted only for Covent Garden. The chief male character, I confess, I should be very unwilling that any one but Kean should play—that is impossible, and I must be contented with an inferior actor. I think you know some of the people of that theatre, or at least, some one who knows them ; and when you have read the play, you may say enough, perhaps, to induce them not to reject it without consideration—but of this, perhaps, if I may judge from the tragedies which they have accepted, there is no danger at any rate.

Write to me as soon as you can on this subject, because it is necessary that I should present it, or, if rejected by the theatre, print it this coming season ; lest somebody else should get hold of it, as the story, which now exists only in manuscript, begins to be generally known among the English. The translation which I send you is to be prefixed to the play, together with a print of Beatrice. I have a copy of her picture by Guido, now in the Colonna palace at Rome —the most beautiful creature you can conceive.

Of course, you will not show the manuscript to any one—and write to me by return of post, at which time the play will be ready to be sent.

I expect soon to write again, and it shall be a less selfish letter. As to Ollier, I don't know what has been published, or what has arrived at his hands.— My *Prometheus*, though ready, I do not send till I know more.

Ever yours, most faithfully,

P. B. S.

## LETTER VIII

To Leigh Hunt, Esq.

*Livorno, Sept. 8th*, 1819.

My dear Friend,—At length has arrived Ollier's parcel, and with it the portrait. What a delightful present ! It is almost yourself, and we sat talking with it, and of it, all the evening. It is a great pleasure to us to possess it, a pleasure in time of need, coming to us when there are few others. How we wish it were you, and not your picture! How I wish we were with you!

This parcel, you know, and all its letters, are now a year old—some older. There are all kinds of dates, from March to August, and 'your date', to use Shakespeare's expression, 'is better in a pie or a pudding, than in your letter'.—'Virginity', Parolles says, but letters are the same thing in another shape.

With it came, too, Lamb's works. I have looked at none of the other books yet. What a lovely thing is his *Rosamund Gray*! How much knowledge of the sweetest and deepest parts of our nature in it! When I think of such a mind as Lamb's—when I see how unnoticed remain things of such exquisite and complete perfection, what should I hope for myself, if I had not higher objects in view than fame?

I have seen too little of Italy, and of pictures. Perhaps P. has shown you some of my letters to him. But at Rome I was very ill, seldom able to go out without a carriage; and though I kept horses for two months there, yet there is so much to see! Perhaps I attended more to sculpture than painting, its forms being more easily intelligible than that of the latter. Yet, I saw the famous works of Raffaele, whom I agree with the whole world in thinking the finest painter. With respect to Michael Angelo I dissent, and think with astonishment and indignation of the common notion that he equals, and, in some respects, exceeds Raffaele. He seems to me to have no sense of moral dignity and loveliness; and the energy for which he has been so much praised, appears to me to be a certain rude, external, mechanical quality, in comparison with anything possessed by Raffaele, or even much inferior artists. His famous painting in the Sixtine Chapel seems to me deficient in beauty and majesty, both in the conception and the execution. He has been called the Dante of painting; but if we find some of the gross and strong outlines which are employed in the most distasteful passages of the *Inferno*, where shall we find *your* Francesca—where the spirit coming over the sea in a boat, like Mars rising from the vapours of the horizon—where Matilda gathering flowers, and all the exquisite tenderness, and sensibility, and ideal beauty, in which Dante excelled all poets except Shakespeare?

As to Michael Angelo's 'Moses'—but you have a cast of that in England. I write these things, heaven knows why!

I have written something and finished it, different from anything else, and a new attempt for me; and I mean to dedicate it to you. I should not have done so without your approbation, but I asked your picture last night, and it smiled assent. If I did not think it in some degree worthy of you, I would not make you a public offering of it. I expect to have to write to you soon about it. If Ollier is not turned Jew, Christian, or become infected with *the Murrain*, he will publish it. Don't let him be frightened, for it is nothing which, by any courtesy of language, can be termed either moral or immoral.

Mary has written to Marianne for a parcel, in which I beg you will make Ollier enclose what you know would most interest me—your *Calendar*, (a sweet extract from which I saw in the *Examiner*,) and the other poems belonging to you; and, for some friends of mine, my *Eclogue*. This parcel, which must be sent instantly, will reach me by October, but don't trust letters to it, except just a line or so. When you write, write by the post.

Ever your affectionate

P. B. S.

## LETTER IX

### TO THOMAS LOVE PEACOCK.

*Leghorn, September* 21*st*, 1819.

MY DEAR PEACOCK,—You will have received a short letter sent with the tragedy, and the tragedy itself by this time. I am, you may believe, anxious to hear what you think of it, and how the manager talks about it. I have printed in Italy 250 copies, because it costs, with all duties and freightage, about half what it would cost in London, and these copies will be sent by sea. My other reason was a belief that the seeing it in print would enable the people at the theatre to judge more easily. Since I last wrote to you, Mr. Gisborne is gone to England for the purpose of obtaining a situation for Henry Reveley. I have given him a letter to you, and you would oblige me by showing him what civilities you can, and by forwarding his views, either by advice or recommendation, as you may find opportunity, not for his sake, who is a great bore, but for the sake of Mrs. Gisborne and Henry Reveley, people for whom we have a great esteem. Henry is a most amiable person, and has great talents as a mechanic and engineer. I have given him also a letter to Hunt, so that you will meet him there. This Mr. Gisborne is a man who knows I cannot tell how many languages, and has read almost all the books you can think of; but all that they contain seems to be to his mind what water is to a sieve. His liberal opinions are all the reflections of Mrs. Gisborne's, a very amiable, accomplished, and completely unprejudiced woman.

Charles Clairmont is now with us on his way to Vienna. He has spent a year or more in Spain, where he has learnt Spanish, and I make him read Spanish all day long. It is a most powerful and expressive language, and I have already learnt sufficient to read with great ease their poet Calderon. I have read about twelve of his plays. Some of them certainly deserve to be ranked amongst the grandest and most perfect productions of the human mind. He exceeds all modern dramatists, with the exception of Shakespeare, whom he resembles, however, in the depth of thought and subtlety of imagination of his writings, and in the rare power of interweaving delicate and powerful comic traits with the most tragical situations, without diminishing their interest. I rate him far above Beaumont and Fletcher.

I have received all the papers you sent me, and the *Examiners* regularly, perfumed with muriatic acid. What an infernal business this of Manchester! What is to be done? Something assuredly. H. Hunt has behaved, I think, with great spirit and coolness in the whole affair.

I have sent you my *Prometheus*, which I do not wish to be sent to Ollier for publication until I write to that effect. Mr. Gisborne will bring it, as also some volumes of Spenser, and the two last of Herodotus and *Paradise Lost*, which may be put with the others.

If my play should be accepted, don't you think it would excite some interest, and take off the unexpected horror of the story, by showing that the events are real, if it could be made to appear in some paper in some form?

You will hear from me again shortly, as I send you by sea the *Cenci's* printed, which you will be good enough to keep. Adieu.

Yours most faithfully,

P. B. SHELLEY.

## LETTER X

### To Leigh Hunt, Esq.

*Livorno, Sept. 27th,* 1819.

My dear Friend,—We are now on the point of leaving this place for Florence, where we have taken pleasant apartments for six months, which brings us to the 1st of April, the season at which new flowers and new thoughts spring forth upon the earth and in the mind. What is then our destination is yet undecided. I have not yet seen Florence, except as one sees the outside of the streets ; but its *physiognomy* indicates it to be a city which, though the ghost of a republic, yet possesses most amiable qualities. I wish you could meet us there in the spring, and we would try to muster up a 'lièta brigata', which, leaving behind them the pestilence of remembered misfortunes, might act over again the pleasures of the Interlocutors in Boccaccio. I have been lately reading this most divine writer. He is, in a high sense of the word, a poet, and his language has the rhythm and harmony of verse. I think him not equal certainly to Dante or Petrarch, but far superior to Tasso and Ariosto, the children of a later and of a colder day. I consider the three first as the productions of the vigour of the infancy of a new nation,—as rivulets from the same spring as that which fed the greatness of the republics of Florence and Pisa, and which checked the influence of the German emperors ; and from which, through obscurer channels, Raffaele and Michael Angelo drew the light and the harmony of their inspiration. When the second-rate poets of Italy wrote, the corrupting blight of tyranny was already hanging on every bud of genius. Energy, and simplicity, and unity of idea, were no more. In vain do we seek in the finest passages of Ariosto and Tasso, any expression which at all approaches in this respect to those of Dante and Petrarch. How much do I admire Boccaccio! What descriptions of nature are those in his little introductions to every new day! It is the morning of life stripped of that mist of familiarity which makes it obscure to us. Boccaccio seems to me to have possessed a deep sense of the fair ideal of human life, considered in its social relations. His more serious theories of love agree especially with mine. He often expresses things lightly too, which have serious meanings of a very beautiful kind. He is a moral casuist, the opposite of the Christian, stoical, ready-made, and worldly system of morals. Do you remember one little remark, or rather maxim of his, which might do some good to the common narrow-minded conceptions of love,—'Bocca bacciata non perde ventura; anzi rinnuova, come fa la luna'?

We expect Mary to be confined towards the end of October. The birth of a child will probably retrieve her from some part of her present melancholy depression.

It would give me much pleasure to know Mr. Lloyd. Do you know, when I was in Cumberland, I got Southey to borrow a copy of Berkeley from him, and I remember observing some pencil notes in it, probably written by Lloyd, which I thought particularly acute. One, especially, struck me as being the assertion of a doctrine, of which even then I had long been persuaded, and on which I had founded much of my persuasions, as regarded the imagined cause of the universe—'Mind cannot create, it can only perceive.' Ask him if he remembers having written it. Of Lamb you know my opinion, and you can bear witness to the regret which I felt, when I learned that the calumny of an enemy had deprived me of his society whilst in England.—Ollier told me that the *Quarterly* are going to review me. I suppose it will be a pretty , and as I am acquiring a taste for humour and drollery, I confess I am curious to see it. I have sent my *Prometheus Unbound* to P.; if you ask him for it, he will show it you. I think it will please you.

Whilst I went to Florence, Mary wrote, but I did not see her letter.—Well, good b'ye. Next Monday I shall write to you from Florence. Love to all.

Most affectionately your friend,

P. B. S.

## LETTER XI

### To Mrs. Gisborne.

*Florence, Nov.* 16, 1819.

Madonna,—I have been lately voyaging in a sea without my pilot, and although my sail has often been torn, my boat become leaky, and the log lost, I have yet sailed in a kind of way from island to island; some of craggy and mountainous magnificence, some clothed with moss and flowers, and radiant with fountains, some barren deserts. I *have been reading Calderon without you.* I have read the *Cisma de Inglaterra*, the *Cabellos de Absolom*, and three or four others. These pieces, inferior to those we read, at least to the *Principe Constante*, in the splendour of particular passages, are perhaps superior in their satisfying completenesss. The *Cabellos de Absolom* is full of the deepest and tenderest touches of nature. Nothing can be more pathetically conceived than the character of old David, and the tender and impartial love, overcoming all insults and all crimes, with which he regards his conflicting and disobedient sons. The incest scene of Amnon and Tamar is perfectly tremendous. Well may Calderon say in the person of the former—

Si sangre sin fuego hiere,
que fara sangre con fuego ?

Incest is, like many other incorrect things, a very poetical circumstance. It may be the excess of love or hate. It may be the defiance of everything for the sake of another, which clothes itself in the glory of the highest heroism; or it may be that cynical rage which, confounding the good and the bad in existing opinions, breaks through them for the purpose of rioting in selfishness and antipathy. Calderon, following the Jewish historians, has represented Amnon's action in the basest point of view—he is a prejudiced savage, acting what he abhors, and abhorring that which is the unwilling party to his crime.

Adieu. Madonna, yours truly,

P. B. S.

I transcribe you a passage from the *Cisma de Inglaterra*—spoken by 'Carlos, Embaxador de Francia, enamorado de Ana Bolena'. Is there anything in Petrarch finer than the second stanza ?[1]

[1] Porque apenas el Sol se coronaba
de nueva luz en la estacion primeva,
quando yo en sus umbrales adoraba
segundo Sol en abreviada esfera;
la noche apenas tremula baxaba,
á solos mis deseos lisonjera,
quando un jardin, republica de flores,
era tercero fiel de mis amores.

Allí, el silencio de la noche fria,
el jazmin, que en las redes se enlazava,
el cristal de la fuente que corria,
el arroyo que á solas murmurava,

## LETTER XII

### To JOHN GISBORNE, Esq.

*Florence, Nov. 16th,* 1819.

MY DEAR SIR,—I envy you the first reading of Theocritus. Were not the Greeks a glorious people? What is there, as Job says of the Leviathan, like unto them? If the army of Nicias had not been defeated under the walls of Syracuse; if the Athenians had, acquiring Sicily, held the balance between Rome and Carthage, sent garrisons to the Greek colonies in the south of Italy, Rome might have been all that its intellectual condition entitled it to be, a tributary, not the conqueror of Greece; the Macedonian power would never have attained to the dictatorship of the civilized states of the world. Who knows whether, under the steady progress which philosophy and social institutions would have made, (for, in the age to which I refer, their progress was both rapid and

el viento que en las hojas se movia,
el Aura que en las flores respirava;
todo era amor; qué mucho, si en tal calma,
aves, fuentes, y flores tienen alma!

No has visto providente y oficiosa
mover el ayre iluminada aveja,
que hasta beber la purpura á la rosa
ya se acerca cobarde, y ya se alexa?
No has visto enamorada mariposa
dar cercos á la luz, hasta que dexa
en monumento facil abrasadas
las alas de color tornasoladas?

Assí mi amor, cobarde muchos dias,
tornos hizo á la rosa y á la llama;
temor che ha sido entre cenizas frias
tantas vezes llorado de quien ama;
pero el amor, que vence con porfias,
y la ocasion, que con disculpas llama,
me animaron, y aveja y mariposa
quemé las alas, y llegué á la rosa.

secure) among a people of the most perfect physical organization—whether the Christian religion would have arisen, or the barbarians have overwhelmed the wrecks of civilization which had survived the conquest and tyranny of the Romans ? What then should we have been ? As it is, all of us who are worth anything, spend our manhood in unlearning the follies, or expiating the mistakes, of our youth. We are stuffed full of prejudices ; and our natural passions are so managed, that if we restrain them we grow intolerant and precise, because we restrain them not according to reason, but according to error ; and if we do not restrain them, we do all sorts of mischief to ourselves and others. Our imagination and understanding are alike subjected to rules the most absurd ; —so much for Theocritus and the Greeks.

In spite of all your arguments, I wish your money were out of the funds. This middle course which you speak of, and which may probably have place, will amount to your losing not all your income, nor retaining all, but have the half taken away. I feel intimately persuaded, whatever political forms may have place in England, that no party can continue many years, perhaps not many months, in the administration, without diminishing the interest of the national debt.—And once having commenced—and having done so safely—where will it end ?

Give Henry my kindest thanks for his most interesting letter, and bid him expect one from me by the next post.

Mary and the babe continue well.—Last night we had a magnificent thunder storm, with claps that shook the house like an earthquake. Both Mary and C—— unite with me in kindest remembrances to all.

Most faithfully yours obliged,

P. B. S.

## LETTER XIII

### To Leigh Hunt, Esq.

*Florence, November*, 1819.

My dear Friend,—Two letters, both bearing date Oct. 20, arrive on the same day; one is always glad of twins.

We hear of a box arrived at Genoa with books and clothes; it must be yours. Meanwhile the babe is wrapt in flannel petticoats, and we get on with him as we can. He is small, healthy, and pretty. Mary is recovering rapidly. Marianne, I hope, is quite well.

You do not tell me whether you have received my lines on the Manchester affair. They are of the exoteric species, and are meant, not for the *Indicator*, but the *Examiner*. I would send for the former, if you like, some letters on such subjects of art as suggest themselves in Italy. Perhaps I will, at a venture, send you a specimen of what I mean next post. I enclose you in this a piece for the *Examiner*, or let it share the fate, whatever that fate may be, of the *Masque of Anarchy*.

I am sorry to hear that you have employed yourself in translating the *Aminta*, though I doubt not it will be a just and beautiful translation. You ought to write Amintas. You ought to exercise your fancy in the perpetual creation of new forms of gentleness and beauty.

With respect to translation, even *I* will not be seduced by it; although the Greek plays, and some of the ideal dramas of Calderon (with which I have lately, and with inexpressible wonder and delight, become acquainted), are perpetually tempting me to throw over their perfect and glowing forms the grey veil of my own words. And you know me too well to suspect that I refrain from a belief that what I could substitute for them would deserve the regret which yours would, if suppressed. I have confidence in my moral sense alone; but that is a kind of originality. I have only translated the *Cyclops* of Euripides, when I could absolutely do nothing else; and the *Symposium* of Plato, which is the delight and astonishment of all who read it; I mean the original, or so much of the original as is seen in my translation, not the translation itself.

I think I have had an accession of strength since my residence in Italy, though the disease itself in the side, whatever it may be, is not subdued. Some day we shall all return from Italy. I fear that in England things will be carried violently by the rulers, and they will not have learned to yield in time to the spirit of the age. The great thing to do is to hold the balance between popular impatience and tyrannical obstinacy; to inculcate with fervour both the right of resistance and the duty of forbearance. You know my principles incite me to take all the good I can get in politics, for ever aspiring to something more. I am one of those whom nothing will fully satisfy, but who are ready to be partially satisfied in all that is practicable. We shall see.

Give Bessy a thousand thanks from me for writing out in that pretty neat hand your kind and powerful defence. Ask what she would like best from Italian land. We mean to bring you all something; and Mary and I have been wondering what it shall be. Do you, each of you, choose.

Adieu, my dear friend.

Yours affectionately ever,

P. B. S.

## LETTER XIV

To Thomas Medwin.

. . . . . . .

*no antidote could know.*

Suppose you erase line 24 which seems superfluous, as one does not see why Oswald shunned the *chase* in particular.

So—you will put in what you think are amendments, and which I have proposed because they appeared such to me. The poem is certainly very beautiful. I think the conclusion rather morbid; that a man should kill himself is one thing, but that he should live on in the dismal way that poor Oswald does, is too much. But it is the spirit of the age, and we are all infected with it.———Send me as soon as you can copies of your printed poems.

*I* have just published a tragedy called the *Cenci* and I see they have reprinted it at Paris at Galignani's. I dare say you will see the French edition, full of errors of course, at Geneva. The people from England tell me it is liked. It is dismal enough. My chief endeavour was to produce a delineation of passions which I had never participated in, in chaste language, and according to the rules of enlightened art.—I don't think very much of it; but it is for you to judge.

Particularly, my dear friend, write to me an account of your motions and when and where we may expect to see you. Are you not tempted by the Baths of Lucca?

I have been seriously ill since I last wrote to you, but I am now recovering.

Affectionately yours,

P. B. S.

*Pisa, May* 1*st* [1820].

## LETTER XV

### To Edmund Ollier.

*Pisa*, 14*th May*, 1820.

Dear Sir,—I reply to your letter by return of post, to confirm what I said in a former letter respecting a new edition of the *Cenci*, which ought, by all means, to be instantly urged forward.

I see by your account that I have been greatly mistaken in my calculations of the profit of my writings. As to the trifle due to me, it may as well remain in your hands.

As to the printing of the *Prometheus*, be it as you will. But, in this case, I shall repose on trust in your care respecting the correction of the press; especially in the lyrical parts, where a minute error would be of much consequence. Mr. Gisborne will revise it; he heard it recited, and will therefore more readily seize any error.

If I had even intended to publish *Julian and Maddalo* with my name, yet I would not print it with *Prometheus*. It would not harmonize. It is an attempt in a different style, in which I am not yet sure of myself, a *sermo pedestris* way of treating human nature, quite opposed to the idealisms of that drama. If you print *Julian and Maddalo*, I wish it to be printed in some unostentatious form, accompanied with the fragments of *Athanase*, and exactly in the manner in which I sent it; and I particularly desire that my name may not be annexed to the first edition of it, in any case.

If *Peter Bell* be printed—you can best judge if it will sell or no, and there would be no other reason for printing such a trifle—attend, I pray you, particularly to completely concealing the author; and for Emma read Betty, as the name of Peter's sister. Emma, I recollect, is the real name of a sister of a great poet who might be mistaken for Peter. I ought to say that I send you poems in a few posts, to print at the end of *Prometheus*, better fitted for that purpose than any in your possession.

Keats, I hope, is going to show himself a great poet; like the sun to burst through the clouds, which, though dyed in the finest colours of the air, obscured his rising. The Gisbornes will bring me from you copies of whatever may be published when they leave England.

Dear Sir,
Yours faithfully,
P. B. SHELLEY.

## LETTER XVI

TO THE EDITOR OF THE 'QUARTERLY REVIEW.'

SIR,—Should you cast your eye on the signature of this letter before you read the contents, you might imagine that they related to a slanderous paper which appeared in your *Review* some time since. I never notice anonymous attacks. The wretch who wrote it has doubtless the additional reward of a consciousness of his motives, besides the thirty guineas a sheet, or whatever it is that you pay him. Of course you cannot be answerable for all the writings which you edit, and *I* certainly bear you no ill-will for having edited the abuse to which I allude—indeed, I was too much amused by being compared to Pharaoh, not readily to forgive editor, printer, publisher, stitcher, or any one, except the despicable writer, connected with something so exquisitely entertaining.

Seriously speaking, I am not in the habit of permitting myself to be disturbed by what is said or written of me, though, I dare say, I may be condemned sometimes justly enough. But I feel, in respect to the writer in question, that 'I am there sitting, where he durst not soar'.

The case is different with the unfortunate subject of this letter, the author of *Endymion*, to whose feelings and situation I entreat you to allow me to call your attention. I write considerably in the dark; but if it is Mr. Gifford that I am addressing, I am persuaded that in an appeal to his humanity and justice, he will acknowledge the *fas ab hoste doceri*. I am aware that the first duty of a Reviewer is towards the public, and I am willing to confess that the *Endymion* is a poem considerably defective, and that, perhaps, it deserved as much censure as the pages of your *Review* record against it; but, not to mention that there is a certain contemptuousness of phraseology from which it is difficult for a critic to abstain, in the review of *Endymion*, I do not think that the writer has given it its due praise. Surely the poem, with all its faults, is a very remarkable production for a man of Keats's age, and the promise of ultimate excellence is such as has rarely been afforded even by such as have afterwards attained high literary eminence. Look at book ii. line 833, &c., and book iii. line 113 to 120—read down that page, and then again from line 193. I could cite many other passages, to convince you that it deserved milder usage. Why it should have been reviewed at all, excepting for the purpose of bringing its excellences into notice, I cannot conceive, for it was very little read, and there was no danger that it should become a model to the age of that false taste, with which I confess that it is replenished.

Poor Keats was thrown into a dreadful state of mind by this review, which, I am persuaded, was not written with any intention of producing the effect, to which it has, at least, greatly contributed, of embittering his existence, and inducing a disease from which there are now but faint hopes of his recovery. The first effects are described to me to have resembled insanity, and it was by assiduous watching that he was restrained from effecting purposes of suicide. The agony of his sufferings at length produced the rupture of a blood-vessel in the lungs, and the usual process of consumption appears to have begun. He is coming to pay me a visit in Italy; but I fear that unless his mind can be kept tranquil, little is to be hoped from the mere influence of climate.

But let me not extort anything from your pity. I have just seen a second volume, published by him evidently in careless despair. I have desired my bookseller to send you a copy, and allow me to solicit your especial attention to the fragment of a poem entitled *Hyperion*, the composition of which was checked by the Review in question. The great proportion of this piece is surely in the very highest style of poetry. I speak impartially, for the canons of taste to which Keats has conformed in his other compositions are the very reverse of my own. I leave you to judge for yourself: it would be an insult to you to suppose that from motives, however honourable, you would lend yourself to a deception of the public.

. . . . . .

(*This letter was never sent.*)

## LETTER XVII

To Thomas Love Peacock.

*Pisa, November (probably 8th)*, 1820.

My dear Peacock,—I also delayed to answer your last letter, because I was waiting for something to say: or at least, something that should be likely to be interesting to you. The box containing my books, and consequently your Essay against the cultivation of poetry, has not arrived; my wonder, meanwhile, in what manner you support such a heresy in this matter-of-fact and money-loving age, holds me in suspense. Thank you for your kindness in correcting *Prometheus*, which I am afraid gave you a great deal of trouble. Among the modern things which have reached me is a volume of poems by Keats: in other respects insignificant enough, but containing the fragment of a poem called *Hyperion*. I dare say you have not time to read it; but it is certainly an astonishing piece of writing, and gives me a conception of Keats which I confess I had not before.

I hear from Mr. Gisborne that you are surrounded with statements and accounts,—a chaos of which you are the God; a sepulchre which encloses in a dormant state the Chrysalis of the Pavonian Psyche. May you start into life some day, and give us another *Melincourt*. Your *Melincourt* is exceedingly admired, and I think much more so than any of your other writings. In this respect the world judges rightly. There is more of the true spirit, and an object less indefinite, than in either *Headlong Hall* or Scythrop.

I am, speaking literally, infirm of purpose. I have great designs, and feeble hopes of ever accomplishing them. I read books, and, though I am ignorant enough, they seem to teach me nothing. To be sure, the reception the public have given me might go far enough to damp any man's enthusiasm. They teach you, it may be said, only what is true. Very true, I doubt not, and the more true the less agreeable. I can compare my experience in this respect to nothing but a series of wet blankets. I have been reading nothing but Greek and Spanish. Plato and Calderon have been my gods. We are now in the town of Pisa. A schoolfellow of mine from India is staying with me, and we are beginning Arabic together. Mary is writing a novel, illustrative of the manners of the Middle Ages in Italy, which she has raked out of fifty old books. I promise myself success from it; and certainly, if what is wholly original will succeed, I shall not be disappointed. . . .

Adieu. *In publica commoda peccem, si longo sermone.*

Ever faithfully yours,

P. B. Shelley.

## LETTER XVIII

To John Gisborne, Esq. (at Leghorn).

*Pisa, oggi,* (*November*, 1820).

My dear Sir,—I send you the *Phaedon* and *Tacitus*. I congratulate you on your conquest of the *Iliad*. You must have been astonished at the perpetually increasing magnificence of the last seven books. Homer there truly begins to be himself. The battle of the Scamander, the funeral of Patroclus, and the high and solemn close of the whole bloody tale in tenderness and inexpiable sorrow, are wrought in a manner incomparable with anything of the same kind. The *Odyssey* is sweet, but there is nothing like this.

*I* am bathing myself in the light and odour of the flowery and starry *Autos.* I have read them all more than once. Henry will tell you how much I am in love with Pacchiani. I suffer from my disease considerably. Henry will also tell you, how much, and how whimsically, he alarmed me last night.

My kindest remembrances to Mrs. Gisborne, and best wishes for your health and happiness.

Faithfully yours,

P. B. S.

I have a new Calderon coming from Paris.

## LETTER XIX

### To Thomas Love Peacock.

*Pisa, February* 15*th,* 1821.

My dear Peacock,—The last letter I received from you, nearly four months from the date thereof, reached me by the boxes which the Gisbornes sent by sea. I am happy to learn that you continue in good external and internal preservation. I received at the same time your printed denunciations against general, and your written ones against particular, poetry; and I agree with you as decidedly in the latter as I differ in the former. The man whose critical gall is not stirred up by such ottava rimas as Barry Cornwall's, may safely be conjectured to possess no gall at all. The world is pale with the sickness of such stuff. At the same time, your anathemas against poetry itself excited me to a sacred rage, or *caloëthes scribendi* of vindicating the insulted Muses. I had the greatest possible desire to break a lance with you, within the lists of a magazine, in honour of my mistress Urania; but God willed that I should be too lazy, and wrested the victory from your hope: since first having unhorsed poetry, and the universal sense of the wisest in all ages, an easy conquest would have remained to you in me, the knight of the shield of shadow and the lance of gossamere. Besides, I was at that moment reading Plato's *Ion*, which I recommend you to reconsider. Perhaps in the comparison of Platonic and Malthusian doctrines, the *mavis errare* of Cicero is a justifiable argument; but I have a whole quiver of arguments on such a subject.

Have you seen Godwin's answer to the apostle of the rich? And what do you think of it? It has not yet reached me, nor has your box, of which I am in daily expectation.

We are now in the crisis and point of expectation in Italy. The Neapolitan and Austrian armies are rapidly approaching each other, and every day the news of a battle may be expected. The former have advanced into the ecclesiastical States, and taken hostages from Rome to assure themselves of the neutrality of that power, and appear determined to try their strength in open battle. I need not tell you how little chance there is that the new and undisciplined levies of Naples should stand against a superior force of veteran troops. But the birth of liberty in nations abounds in examples of a reversal of the ordinary laws of calculation: the defeat of the Austrians would be the signal of insurrection throughout all Italy.

I am devising literary plans of some magnitude. But nothing is more difficult and unwelcome than to write without a confidence of finding readers; and if my play of the *Cenci* found none or few, I despair of ever producing anything that shall merit them.

Among your anathemas of the modern attempts in poetry, do you include Keats's *Hyperion*? I think it very fine. His other poems are worth little; but if the *Hyperion* be not grand poetry, none has been produced by our contemporaries.

I suppose *you* are writing nothing but Indian laws, &c. I have but a faint idea of your occupation; but I suppose it has much to do with pen and ink.

Mary desires to be kindly remembered to you; and I remain, my dear Peacock, yours very faithfully,

P. B. SHELLEY.

## LETTER XX

TO THOMAS LOVE PEACOCK.

*Pisa, March* 21*st*, 1821.

MY DEAR PEACOCK,—I dispatch by this post the first part of an essay intended to consist of three parts, which I design for an antidote to your *Four Ages of Poetry*. You will see that I have taken a more general view of what is poetry than you have, and will perhaps agree with several of my positions, without considering your own touched. But read and judge; and do not let us imitate the great founders of the picturesque, Price and Payne Knight, who, like two ill-trained beagles, began snarling at each other when they could not catch the hare.

I hear the welcome news of a box from England announced by Mr. Gisborne. How much new poetry does it contain? The Bavii and Maevii of the day are very fertile; and I wish those who honour me with boxes would read and inwardly digest your *Four Ages of Poetry*; for I had much rather, for my own private reading, receive political, geological, and moral treatises than this stuff in *terza*, *ottava*, and *tremillesima rima* whose earthly baseness has attracted the lightning of your undiscriminating censure upon the temple of immortal song. Procter's verses enrage me far more than those of Codrus did Juvenal, and with better reason. Juvenal need not have been stunned unless he had liked it; but my boxes are packed with this trash, to the exclusion of what I want to see. But your box will make amends.

Do you see much of Hogg now? and the Boinvilles and Colson? Hunt I suppose not. And are you occupied as much as ever? We are surrounded here in Pisa by revolutionary volcanoes, which, as yet, give more light than heat; the lava has not yet reached Tuscany. But the news in the papers will tell you far more than it is prudent for me to say; and for this once I will observe your rule of political silence. The Austrians wish that the Neapolitans and Piedmontese would do the same.

We have seen a few more people than usual this winter, and have made a very interesting acquaintance with a Greek Prince, perfectly acquainted with ancient literature, and full of enthusiasm for the liberties and improvement of his country. Mary has been a Greek student for several months, and is reading *Antigone* with our turbaned friend, who, in return, is taught English. Claire has passed the Carnival at Florence, and has been preternaturally gay. I have had a severe ophthalmia, and have read or written little this winter; and have made acquaintance in an obscure convent with the only Italian for whom I ever felt any interest.

I want you to do something for me: that is, to get me two pounds' worth of Tassi's gems, in Leicester Square, the prettiest, according to your taste; among them, the head of Alexander; and to get me two seals engraved and set, one smaller, and the other handsomer; the device a dove with outspread wings, and this motto round it:

Μάντις εἰμ' ἐσθλῶν ἀγώνων.

Mary desires her best regards; and I remain, my dear Peacock, ever most sincerely yours,

P. B. S.

## LETTER XXI

### To Mr. and Mrs. Gisborne.

*Bagni, Tuesday Evening*, (*June 5th*, 1821).

My dear Friends,—We anxiously expect your arrival at the Baths; but as I am persuaded that you will spend as much time with us as you can save from your necessary occupations before your departure, I will forbear to vex you with importunity. My health does not permit me to spend many hours from home. I have been engaged these last days in composing a poem on the death of Keats, which will shortly be finished; and I anticipate the pleasure of reading it to you, as some of the very few persons who will be interested in it and understand it. It is a highly-wrought *piece of art*, and perhaps better, in point of composition, than anything I have written.

I have obtained a purchaser for some of the articles of your three lists, a catalogue of which I subjoin. I shall do my utmost to get more; could you not send me a complete list of your *furniture*, as I have had inquiries made about chests of drawers, &c.

My unfortunate box! it contained a chaos of the elements of *Charles I*. If the idea of the *creator* had been packed up with them, it would have shared the same fate; and that, I am afraid, has undergone another sort of shipwreck.

Very faithfully and affectionately yours,

S.

## LETTER XXII

### To John Gisborne, Esq.

*Pisa; Saturday*, (*June* 16*th*, 1821).

My dear Friend,—I have received the heartrending account of the closing scene of the great genius whom envy and ingratitude scourged out of the world. I do not think that if I had seen it before, I could have composed my poem. The enthusiasm of the imagination would have overpowered the sentiment.

As it is, I have finished my *Elegy*; and this day I send it to the press at Pisa. You shall have a copy the moment it is completed. I think it will please you. I have dipped my pen in consuming fire for his destroyers; otherwise the style is calm and solemn.

Pray, when shall we see you? Or are the streams of Helicon less salutary than sea-bathing for the nerves? Give us as much as you can before you go to England, and rather divide the term than not come soon.

Mrs. —— wishes that none of the books, desk, &c., should be packed up with the piano; but that they should be sent, one by one, by Pepi. Address them to *me* at her house. She desired me to have them addressed to *me*, why I know not.

A droll circumstance has occurred. *Queen Mab*, a poem written by me when very young, in the most furious style, with long notes against Jesus Christ, and God the Father, and the king, and bishops, and marriage, and the devil knows what, is just published by one of the low booksellers in the Strand, against my wish and consent, and all the people are at loggerheads about it. H. S. gives me this account. You may imagine how much I am amused. For the sake of a dignified appearance, however, and really because I wish to protest against all the bad poetry in it, I have given orders to say that it is all done against my desire, and have directed my attorney to apply to Chancery for an injunction, which he will not get.

I am pretty ill, I thank you, just now; but I hope you are better.

Most affectionately yours,

P. B. S.

## LETTER XXIII

To Mrs. Shelley.

*Ravenna, August* 7, 1821.

My dearest Mary,—I arrived last night at ten o'clock, and sat up talking with Lord Byron until five this morning. I then went to sleep, and now awake at eleven, and having despatched my breakfast as quick as possible, mean to devote the interval until twelve, when the post departs, to you.

Lord Byron is very well, and was delighted to see me. He has in fact completely recovered his health, and lives a life totally the reverse of that which he led at Venice. He has a permanent sort of liaison with Contessa Guiccioli, who is now at Florence, and seems from her letters to be a very amiable woman. She is waiting there until something shall be decided as to their emigration to Switzerland or stay in Italy; which is yet undetermined on either side. She was compelled to escape from the Papal territory in great haste, as measures had already been taken to place her in a covent, where she would have been unrelentingly confined for life. The oppression of the marriage contract, as existing in the laws and opinions of Italy, though less frequently exercised, is far severer than that of England. I tremble to think of what poor Emilia is destined to.

Lord Byron had almost destroyed himself in Venice: his state of debility was such that he was unable to digest any food, he was consumed by hectic fever, and would speedily have perished, but for this attachment, which has reclaimed him from the excesses into which he threw himself from carelessness and pride, rather than taste. Poor fellow! he is now quite well, and immersed in politics and literature. He has given me a number of the most interesting details on the former subject, but we will not speak of them in a letter. Fletcher is here; and as if, like a shadow, he waxed and waned with the substance of his master: Fletcher also has recovered his good looks, and from amidst the unseasonable grey hairs, a fresh harvest of flaxen locks put forth.

We talked a great deal of poetry, and such matters last night; and as usual differed, and I think more than ever. He affects to patronize a system of criticism fit for the production of mediocrity, and although all his fine poems and passages have been produced in defiance of this system, yet I recognize the pernicious effects of it in the *Doge of Venice*; and it will cramp and limit his future efforts however great they may be, unless he gets rid of it. I have read only parts of it, or rather he himself read them to me, and gave me the plan of the whole.

Lord Byron has also told me of a circumstance that shocks me exceedingly; because it exhibits a degree of desperate and wicked malice for which I am at a loss to account. When I hear such things my patience and my philosophy are put to a severe proof, whilst I refrain from seeking out some obscure hiding-place, where the countenance of man may never meet me more.

. . . Imagine my despair of good, imagine how is it possible that one of so weak and sensitive a nature as mine can run further the gauntlet through this hellish society of men. *You* should write to the Hoppners a letter refuting the charge, in case you believe, and know, and can prove that it is false; stating the grounds and proofs of your belief. I need not dictate what you should say; nor, I hope, inspire you with warmth to rebut a charge, which you only can effectually rebut. If you will send the letter to me here, I will forward it to the Hoppners. Lord Byron is not up, I do not know the Hoppners' address, and I am anxious not to lose a post.

## LETTER XXIV

### To Mrs. Shelley.

*Friday*, (*August* 9, 1821).

We ride out in the evening, through the pine forests which divide this city from the sea. Our way of life is this, and I have accommodated myself to it without much difficulty:—L. B. gets up at two, breakfasts; we talk, read, &c., until six; then we ride, and dine at eight; and after dinner sit talking till four or five in the morning. I get up at twelve, and am now devoting the interval between my rising and his, to you.

L. B. is greatly improved in every respect. In genius, in temper, in moral views, in health, in happiness. The connexion with la Guiccioli has been an inestimable benefit to him. He lives in considerable splendour, but within his income, which is now about 4000*l.* a-year; 100*l.* of which he devotes to purposes of charity. He has had mischievous passions, but these he seems to have subdued, and he is becoming what he should be, a virtuous man. The interest which he took in the politics of Italy, and the actions he performed in consequence of it, are subjects not fit to be *written*, but are such as will delight and surprise you. He is not yet decided to go to Switzerland—a place, indeed, little fitted for him: the gossip and the cabals of those anglicised coteries would torment him, as they did before, and might exasperate him into a relapse of libertinism, which he says he plunged into not from taste, but despair. La Guiccioli and her brother (who is L. B's friend and confidant, and acquiesces perfectly in her connexion with him,) wish to go to Switzerland, as L. B. says, merely from the novelty of the pleasure of travelling. L. B. prefers Tuscany or Lucca, and is trying to persuade them to adopt his views. He has made *me* write a long letter to her to engage her to remain—an odd thing enough for an utter stranger to write on subjects of the utmost delicacy to his friend's mistress. But it seems destined that I am always to have some active part in every body's affairs whom I approach. I have set down in lame Italian, the strongest reasons I can think of against the Swiss emigration—to tell you truth, I should be very glad to accept, as my fee, his establishment in Tuscany. Ravenna is a miserable place; the people are barbarous and wild, and their language the most infernal *patois* that you can imagine. He would be, in every respect, better among the Tuscans. I am afraid he would not like Florence, on account of the English there. There is Lucca,

Florence, Pisa, Siena, and I think nothing more. What think you of Prato, or Pistoia, for him ?—no Englishman approaches those towns ; but I am afraid no house could be found good enough for him in that region.

He has read to me one of the unpublished cantos of *Don Juan*, which is astonishingly fine. It sets him not only above, but far above, all the poets of the day —every word is stamped with immortality. I despair of rivalling Lord Byron, as well I may, and there is no other with whom it is worth contending. This canto is in the style, but totally, and sustained with incredible ease and power, like the end of the second canto. There is not a word which the most rigid assertor of the dignity of human nature would desire to be cancelled. It fulfils, in a certain degree, what I have long preached of producing—something wholly new and relative to the age, and yet surpassingly beautiful. It may be vanity, but I think I see the trace of my earnest exhortations to him to create something wholly new. He has finished his *life* up to the present time, and given it to Moore, with liberty for Moore to sell it for the best price he can get, with condition that the bookseller should publish it after his death. Moore has sold it to Murray for *two thousand pounds*. I have spoken to him of Hunt, but not with a direct view of demanding a contribution ; and, though I am sure that if asked it would not be refused—yet there is something in me that makes it impossible. Lord Byron and I are excellent friends, and were I reduced to poverty, or were I a writer who had no claims to a higher station than I possess—or did I possess a higher than I deserve, we should appear in all things as such, and I would freely ask him any favour. Such is not the case. The demon of mistrust and pride lurks between two persons in our situation, poisoning the freedom of our intercourse. This is a tax, and a heavy one, which we must pay for being human. I think the fault is not on my side, nor is it likely, I being the weaker. I hope that in the next world these things will be better managed. What is passing in the heart of another, rarely escapes the observation of one who is a strict anatomist of his own.

## LETTER XXV

### To Thomas Love Peacock.

*Ravenna, August (probably 10th)*, 1821.

My dear Peacock,—I received your last letter just as I was setting off from the Bagni on a visit to Lord Byron at this place. Many thanks for all your kind attention to my accursed affairs. . . .

I have sent you by the Gisbornes a copy of the *Elegy on Keats*. The subject, I know, will not please you; but the composition of the poetry, and the taste in which it is written, I do not think bad. You and the enlightened public will judge. Lord Byron is in excellent cue both of health and spirits. He has got rid of all those melancholy and degrading habits which he indulged at Venice. He lives with one woman, a lady of rank here, to whom he is attached, and who is attached to him, and is in every respect an altered man. He has written three more cantos of *Don Juan*. I have yet only heard the fifth, and I think that every word of it is pregnant with immortality. I have not seen his late plays, except *Marino Faliero*, which is very well, but not so transcendently fine as the *Don Juan*. Lord Byron gets up at two. I get up, quite contrary to my usual custom (but one must sleep or die, like Southey's sea-snake in *Kehama*), at twelve. After breakfast, we sit talking till six. From six till eight we gallop through the pine forests which divide Ravenna from the sea; we then come home and dine, and sit up gossiping till six in the morning. I don't suppose this will kill me in a week or fortnight, but I shall not try it longer. Lord B.'s establishment consists, besides servants, of ten horses, eight enormous dogs, three monkeys, five cats, an eagle, a crow, and a falcon; and all these, except the horses, walk about the house, which every now and then resounds with their unarbitrated quarrels, as if they were the masters of it. Lord B. thinks you wrote a pamphlet signed *John Bull*; he says he knew it by the style resembling *Melincourt*, of which he is a great admirer. I read it, and assured him that it could not possibly be yours. I write nothing, and probably shall write no more. It offends me to see my name classed among those who have no name. If I cannot be something better, I had rather be nothing. . . . and the accursed cause to the downfall of which I dedicated what powers I may have had—flourishes like a cedar and covers England with its boughs. My motive was never the infirm desire of fame; and if I should continue an author, I feel that I should desire it. This cup is justly given to one only of an age; indeed, participation would make it worthless: and unfortunate they who seek it and find it not.

I congratulate you—I hope I ought to do so—on your expected stranger. He is introduced into a rough world. My regards to Hogg, and Colson if you see him.

Ever most faithfully yours,

P. B. S.

After I have sealed my letter, I find that my enumeration of the animals in this Circaean Palace was defective, and that in a material point. I have just met on the grand staircase five peacocks, two guinea-hens, and an Egyptian crane. I wonder who all these animals were, before they were changed into these shapes.

## LETTER XXVI

### To Mrs. Shelley.

*Saturday—Ravenna.*

My dear Mary,—You will be surprised to hear that L. B. has decided upon coming to *Pisa*, in case he shall be able, with my assistance, to prevail upon his mistress to remain in Italy, of which I think there is little doubt. He wishes for a large and magnificent house, but he has furniture of his own, which he would send from Ravenna. Inquire if any of the large palaces are to be let. We discussed Prato, Pistoia, Lucca, &c., but they would not suit him so well as Pisa, to which, indeed, he shows a decided preference. So let it be! Florence he objects to, on account of the prodigious influx of English.

I don't think this circumstance ought to make any difference in our own plans with respect to this winter in Florence, because we could easily reassume our station, with the spring, at Pugnano or the baths, in order to enjoy the society of the noble lord. But do you consider this point, and write to me your full opinion, at the Florence post-office.

I suffer much to-day from the pain in my side, brought on, I believe, by this accursed water. In other respects, I am pretty well, and my spirits are much improved; they had been improving, indeed, before I left the baths, after the deep dejection of the early part of the year.

I am reading *Anastasius*. One would think that L. B. had taken his idea of the three last cantos of *Don Juan* from this book. That, of course, has nothing to do with the merit of this latter, poetry having nothing to do with the invention of facts. It is a very powerful, and very entertaining novel, and a faithful picture, they say, of modern Greek manners. I have read L. B.'s Letter to Bowles; some good things—but he ought not to write prose criticism.

You will receive a long letter, sent with some of L. B's, express to Florence. I write this in haste.

Yours most affectionately,

S.

## LETTER XXVII

### To Leigh Hunt, Esq.

*Pisa, August 26th*, 1821.

My dearest Friend,—Since I last wrote to you, I have been on a visit to Lord Byron at Ravenna. The result of this visit was a determination, on his part, to come and live at Pisa; and I have taken the finest palace on the Lung' Arno for him. But the material part of my visit consists in a message which he desires me to give you, and which, I think, ought to add to your determination—for such a one I hope you have formed, of restoring your shattered health and spirits by a migration to these 'regions mild of calm and serene air'.

He proposes that you should come and go shares with him and me, in a periodical work, to be conducted here; in which each of the contracting parties should publish all their original compositions, and share the profits. He proposed it to Moore, but for some reason it was never brought to bear. There can be no doubt that the *profits* of any scheme in which you and Lord Byron engage, must, from various, yet co-operating reasons, be very great. As for myself, I am, for the present, only a sort of link between you and him, until you can know each other, and effectuate the arrangement; since (to entrust you with a secret which, for your sake, I withhold from Lord Byron) nothing would induce me to share in the profits, and still less, in the borrowed splendour of such a partnership. You and he, in different manners, would be equal, and would bring, in a different manner, but in the same proportion, equal stocks of reputation and success. Do not let my frankness with you, nor my belief that you deserve it more than Lord Byron, have the effect of deterring you from assuming a station in modern literature, which the universal voice of my contemporaries forbids me either to stoop or to aspire to. I am, and I desire to be, nothing.

I did not ask Lord Byron to assist me in sending a remittance for your journey; because there are men, however excellent, from whom we would never receive an obligation, in the worldly sense of the word; and I am as jealous for my friend as for myself; but I suppose that I shall at last make up an impudent face, and ask Horace Smith to add to the many obligations he has conferred on me. I know I need only ask.

I think I have never told you how very much I like your *Amyntas*; it almost reconciles me to translations. In another sense I still demur. You might have written another such poem as the *Nymphs*, with no great access of efforts. I am full of thoughts and plans, and should do something, if the feeble and irritable frame which encloses it was willing to obey the spirit. I fancy that then I should do great things.

Before this you will have seen *Adonais.* Lord Byron, I suppose from modesty, on account of his being mentioned in it, did not say a word of *Adonais,* though he was loud in his praise of *Prometheus,* and, what you will not agree with him in, censure of the *Cenci.* Certainly, if *Marino Faliero* is a drama, the *Cenci* is not—but that between ourselves. Lord Byron is reformed, as far as gallantry goes, and lives with a beautiful and sentimental Italian lady, who is as much attached to him as may be. I trust greatly to his intercourse with you, for his creed to become as pure as he thinks his conduct is. He has many generous and exalted qualities, but the canker of aristocracy wants to be cut out.

## LETTER XXVIII

### To Horatio Smith, Esq.

*Pisa, Sept.* 14*th,* 1821.

My dear Smith,—I cannot express the pain and disappointment with which I learn the change in your plans, no less than the afflicting cause of it. Florence will no longer have any attractions for me this winter, and I shall contentedly sit down in this humdrum Pisa, and refer to hope and to chance the pleasure I had expected from your society this winter. What shall I do with your packages, which have now, I believe, all arrived at Guebhard's at Leghorn? Is it not possible that a favourable change in Mrs. Smith's health might produce a corresponding change in your determinations, and would it, or would it not, be premature to forward the packages to your present residence, or to London? I will pay every possible attention to your instructions in this regard.

I had marked down several houses in Florence, and one especially on the Arno, a most lovely place, though they asked rather more than perhaps you would have chosen to pay—yet nothing approaching to an English price.—I do not yet entirely give you up.—Indeed, I should be sorry not to hope that Mrs. Smith's state of health would not soon become such, as to remove your principal objection to this delightful climate. I have not, with the exception of three or four days, suffered in the least from the heat this year. Though, it is but fair to confess, that my temperament approaches to that of the salamander.

We expect Lord Byron here in about a fortnight. I have just taken the finest palace in Pisa for him, and his luggage, and his horses, and all his train, are, I believe, already on their way hither. I dare say you have heard of the life he led at Venice, rivalling the wise Solomon, almost, in the number of his concubines. Well, he is now quite reformed, and is leading a most sober and decent life, as *cavaliere servente* to a very pretty Italian woman, who has already arrived at Pisa, with her father and her brother (such are the manners of Italy,) as the jackals of the lion. He is occupied in forming a new drama, and, with views which I doubt not will expand as he proceeds, is determined to write a series of plays, in which he will follow the French tragedians and Alfieri, rather than those of England and Spain, and produce something new at least to England. This seems to me the wrong road; but genius like his is destined to lead and not to follow. He will shake off his shackles as he finds they cramp him. I believe he will produce something very great; and that familiarity with the dramatic power of human nature, will soon enable him to soften down the severe and unharmonizing traits of his *Marino Faliero*. I think you know Lord

Byron personally, or is it your brother? If the latter, I know that he wished particularly to be introduced to you, and that he will sympathize, in some degree, in this great disappointment which I feel in the change, or, as I yet hope, in the prorogation of your plans.

I am glad you like *Adonais* and, particularly, that you do not think it metaphysical, which I was afraid it was. I was resolved to pay some tribute of sympathy to the unhonoured dead, but I wrote, as usual, with a total ignorance of the effect that I should produce.—I have not yet seen your pastoral drama; if you have a copy, could you favour me with it? It will be six months before I shall receive it from England. I have heard it spoken of with high praise, and I have the greatest curiosity to see it.

The Gisbornes promised to buy me some books in Paris, and I had asked you to be kind enough to advance them what they might want to pay for them. I cannot conceive why they did not execute this little commission for me, as they knew how very much I wished to receive these books by the same conveyance as the filtering-stone. Dare I ask you to do me the favour to buy them? *A complete edition of the works of Calderon,* and the French translation of Kant, a German *Faust*, and to add the *Nympholept*?—I am indifferent as to a little more or less expense, so that I may have them immediately. I will send you an order on Paris for the amount, together with the thirty-two francs you were kind enough to pay for me.

All public attention is now centred on the wonderful revolution in Greece. I dare not, after the events of last winter, hope that slaves can become freemen so cheaply; yet I know one Greek of the highest qualities, both of courage and conduct, the Prince Mavrocordato, and if the rest be like him, all will go well.—The news of this moment is, that the Russian army has orders to advance.

Mrs. S. unites with me in the most heartfelt regret,

And I remain, my dear Smith,

Most faithfully yours,

P. B. S.

If you happen to have brought a copy of Clarke's edition of *Queen Mab* for me, I should like very well to see it.—I really hardly know what this poem is about. I am afraid it is rather rough.

## LETTER XXIX

### To John Gisborne, Esq.

*Pisa, October* 22, 1821.

My dear Gisborne,—At length the post brings a welcome letter from you, and I am pleased to be assured of your health and safe arrival. I expect with interest and anxiety the intelligence of your progress in England, and how far the advantages there compensate the loss of Italy. I hear from Hunt that he is determined on emigration, and if I thought the letter would arrive in time, I should beg you to suggest some advice to him. But you ought to be incapable of forgiving me the fact of depriving England of what it must lose when Hunt departs.

Did I tell you that Lord Byron comes to settle at Pisa, and that he has a plan of writing a periodical work in conjunction with Hunt. His house, Madame Felichi's, is already taken and fitted up for him, and he has been expected every day these six weeks. La Guiccioli, who awaits him impatiently, is a very pretty, sentimental, innocent Italian, who has sacrificed an immense fortune for the sake of Lord Byron, and who, if I know anything of my friend, of her and of human nature, will hereafter have plenty of leisure and opportunity to repent her rashness. Lord Byron is, however, quite cured of his gross habits, as far as habits; the perverse ideas on which they were formed, are not yet eradicated.

We have furnished a house at Pisa, and mean to make it our head-quarters. I shall get all my books out, and entrench myself like a spider in a web. If you can assist P. in sending them to Leghorn, you would do me an especial favour; but do not buy me Calderon, *Faust*, or Kant, as H. S. promises to send them me from Paris, where I suppose you had not time to procure them. Any other books you or Henry think would accord with my design, Ollier will furnish you with.

I should like very much to hear what is said of my *Adonais*, and you would oblige me by cutting out, or making Ollier cut out, any respectable criticism on it, and sending it me; you know I do not mind a crown or two in postage. The *Epipsychidion* is a mystery; as to real flesh and blood, you know that I do not deal in those articles; you might as well go to a gin-shop for a leg of mutton, as expect anything human or earthly from me. I desired Ollier not to circulate this piece except to the συνετοί, and even they, it seems, are inclined to approximate me to the circle of a servant girl and her sweetheart. But I intend to write a Symposium of my own to set all this right.

I am just finishing a dramatic poem, called *Hellas*, upon the contest now raging in Greece—a sort of imitation of the *Persae* of Aeschylus, full of lyrical poetry. I try to be what I might have been, but am not successful. I find that (I dare say I shall quote wrong,)—

Dem herrlichsten, den sich der Geist emprägt
Drängt immer fremd und fremder Stoff sich an.

The *Edinburgh* review lies. Godwin's answer to Malthus is victorious and decisive ; and that it should not be generally acknowledged as such, is full evidence of the influence of successful evil and tyranny. What Godwin is, compared to Plato and Lord Bacon, we well know ; but compared with these miserable sciolists, he is a vulture to a worm.

I read the Greek dramatists and Plato for ever. You are right about *Antigone* ; how sublime a picture of a woman ! And what think you of the choruses, and especially the lyrical complaints of the godlike victim ; and the menaces of Tiresias, and their rapid fulfilment ? Some of us have, in a prior existence, been in love with an Antigone, and that makes us find no full content in any mortal tie. As to books, I advise you to live near the British Museum, and read there. I have read, since I saw you, the *Jungfrau von Orleans* of Schiller,—a fine play, if the fifth act did not fall off. Some Greeks, escaped from the defeat in Wallachia, have passed through Pisa to re-embark at Leghorn for the Morea ; and the Tuscan Government allowed them, during their stay and passage, three lire each per day and their lodging ; that is good. Remember me and Mary most kindly to Mrs. Gisborne and Henry, and believe me,

Yours most affectionately,

P. B. S.

## LETTER XXX

### To Joseph Severn.

*Pisa, Nov. 29th,* 1821.

Dear Sir,—I send you the elegy on poor Keats—and I wish it were better worth your acceptance. You will see, by the preface, that it was written before I could obtain any particular account of his last moments; all that I still know, was communicated to me by a friend who had derived his information from Colonel Finch; I have ventured to express, as I felt, the respect and admiration which *your* conduct towards him demands.

In spite of his transcendent genius, Keats never was, nor ever will be, a popular poet; and the total neglect and obscurity in which the astonishing remnants of his mind still lie, was hardly to be dissipated by a writer, who, however he may differ from Keats in more important qualities, at least resembles him in that accidental one, a want of popularity.

I have little hope, therefore, that the poem I send you will excite any attention, nor do I feel assured that a critical notice of his writings would find a single reader. But for these considerations, it had been my intention to have collected the remnants of his compositions, and to have published them with a Life and Criticism. Has he left any poems or writings of whatsoever kind, and in whose possession are they? Perhaps you would oblige me by information on this point.

Many thanks for the picture you promise me: I shall consider it among the most sacred relics of the past.

For my part, I little expected, when I last saw Keats at my friend Leigh Hunt's, that I should survive him.

Should you ever pass through Pisa, I hope to have the pleasure of seeing you, and of cultivating an acquaintance into something pleasant, begun under such melancholy auspices.

Accept, my dear sir, the assurance of my sincere esteem, and believe me,

Your most sincere and faithful servant,

PERCY B. SHELLEY.

Do you know Leigh Hunt? I expect him and his family *here* every day.

## LETTER XXXI

### TO THOMAS LOVE PEACOCK.

*Pisa, January* (*probably* 11*th*), 1822.

MY DEAR PEACOCK,—. . . I am still at Pisa, where I have at length fitted up some rooms at the top of a lofty palace that overlooks the city and the surrounding region, and have collected books and plants about me, and established myself for some indefinite time, which, if I read the future, will not be short. I wish you to send my books by the very first opportunity, and I expect in them a great augmentation of comfort. Lord Byron is established here, and we are constant companions. No small relief this, after the dreary solitude of the understanding and the imagination in which we passed the first years of our expatriation, yoked to all sorts of miseries and discomforts.

Of course you have seen his last volume, and if you before thought him a great poet, what is your opinion now that you have read *Cain*! The *Foscari* and *Sardanapalus* I have not seen; but as they are in the style of his later writings, I doubt not they are very fine. We expect Hunt here every day, and remain in great anxiety on account of the heavy gales which he must have encountered at Christmas. Lord Byron has fitted up the lower apartments of his palace for him, and Hunt will be agreeably surprised to find a commodious lodging prepared for him after the fatigues and dangers of his passage. I have been long idle, and, as far as writing goes, despondent; but I am now engaged on *Charles the First*, and a devil of a nut it is to crack.

Mary and Clara (who is not with us just at present) are well, and so is our little boy, the image of poor William. We live as usual, tranquilly. I get up, or at least wake early; read and write till two; dine; go to Lord B.'s, and ride, or play billiards, as the weather permits; and sacrifice the evening either to light books or whoever happens to drop in. Our furniture, which is very neat, cost fewer shillings than that at Marlow did pounds sterling; and our windows are full of plants, which turn the sunny winter into spring. My health is better—my cares are lighter; and although nothing will cure the consumption of my purse, yet it drags on a sort of life in death, very like its master, and seems, like Fortunatus's, always empty yet never quite exhausted. You will have seen my *Adonais* and perhaps my *Hellas*, and I think, whatever you may judge of the subject, the composition of the first poem will not wholly displease you. I wish I had something better to do than furnish this jingling food for the hunger of oblivion, called verse, but I have not; and since you give me no encouragement about India I cannot hope to have.

How is your little star, and the heaven which contains the milky way in which it glimmers?

Adieu.—Yours ever most truly,

S.

## LETTER XXXII

### To John Gisborne, Esq.

*Pisa, April* 10, 1822.

My dear Gisborne,—I have received *Hellas*, which is prettily printed, and with fewer mistakes than any poem I ever published. Am I to thank you for the revision of the press? or who acted as midwife to this last of my orphans, introducing it to oblivion, and me to my accustomed failure? May the cause it celebrates be more fortunate than either! Tell me how you like *Hellas*, and give me your opinion freely. It was written without much care, and in one of those few moments of enthusiasm which now seldom visit me, and which make me pay dear for their visits. I know what to think of *Adonais*, but what to think of those who confound it with the many bad poems of the day, I know not.

I have been reading over and over again *Faust*, and always with sensations which no other composition excites. It deepens the gloom and augments the rapidity of ideas, and would therefore seem to me an unfit study for any person who is a prey to the reproaches of memory, and the delusions of an imagination not to be restrained. And yet the pleasure of sympathizing with emotions known only to few, although they derive their sole charm from despair, and the scorn of the narrow good we can attain in our present state, seems more than to ease the pain which belongs to them. Perhaps all discontent with the *less* (to use a Platonic sophism,) supposes the sense of a just claim to the *greater*, and that we admirers of *Faust* are on the right road to Paradise. Such a supposition is not more absurd, and is certainly less demoniacal, than that of Wordsworth, where he says—

This earth,
Which is the world of all of us, and where
*We find our happiness, or not at all.*

As if, after sixty years' suffering here, we were to be roasted alive for sixty million more in hell, or charitably annihilated by a *coup de grace* of the bungler who brought us into existence at first!

Have you read Calderon's *Magico Prodigioso*? I find a striking singularity between *Faust* and this drama, and if I were to acknowledge Coleridge's distinction, should say Goethe was the *greatest* philosopher, and Calderon the *greatest* poet. *Cyprian* evidently furnished the *germ* of *Faust*, as *Faust* may furnish the germ of other poems; although it is as different from it in structure and plan as the acorn from the oak. I have—imagine my presumption—translated several scenes from both, as the basis of a paper for our journal. I am well content with those from Calderon, which in fact gave me very little trouble; but those from *Faust*—I feel how imperfect a representation, even with all the licence I assume to figure to myself how Goethe would have written in English, my words convey. No one but Coleridge is capable of this work.

We have seen here a translation of some scenes, and indeed the most remarkable ones, accompanying those astonishing etchings which have been published in England from a German master. It is not bad—and faithful enough—but how weak! how incompetent to represent *Faust*! I have only attempted the scenes omitted in this translation, and would send you that of the *Walpurgisnacht*, if I thought Ollier would place the postage to my account. What etchings those are! I am never satiated with looking at them;

and, I fear, it is the only sort of translation of which *Faust* is susceptible. I never perfectly understood the Hartz Mountain scene, until I saw the etching; and then, Margaret in the summer-house with Faust! The artist makes one envy his happiness that he can sketch such things with calmness, which I only dared look upon once, and which made my brain swim round only to touch the leaf on the opposite side of which I knew that it was figured. Whether it is that the artist has surpassed *Faust*, or that the pencil surpasses language in some subjects, I know not, or that I am more affected by a visible image, but the etching certainly excited me far more than the poem it illustrated. Do you remember the fifty-fourth letter of the first part of the *Nouvelle Héloïse*? Goethe, in a subsequent scene, evidently had that letter in his mind, and this etching is an idealism of it. So much for the world of shadows!

What think you of Lord Byron's last volume? In my opinion it contains finer poetry than has appeared in England since the publication of *Paradise Regained*. *Cain* is apocalyptic; it is a revelation not before communicated to man. I write nothing but by fits. I have done some of *Charles I*; but although the poetry succeeded very well, I cannot seize on the conception of the subject as a whole, and seldom now touch the canvas. You know I don't think much about Reviews, nor of the fame they give, nor that they take away. It is absurd in any Review to criticize *Adonais*, and still more to pretend that the verses are bad. *Prometheus* was never intended for more than five or six persons.

And how are you getting on? Do your plans still want success? Do you regret Italy? or anything that Italy contains? And in case of an entire failure in your expectations, do you think of returning here?

You see the first blow has been made at funded property :—do you intend to confide and invite a second ? You would already have saved something per cent., if you had invested your property in Tuscan land. The next best thing would be to invest it in English, and reside upon it. I tremble for the consequences, to you personally, from a prolonged confidence in the funds. Justice, policy, the hopes of the nation and renewed institutions, demand your ruin, and I, for one, cannot bring myself to desire what is in itself desirable, till you are free. You see how liberal I am of advice ; but you know the motives that suggest it. What is Henry about, and how are his prospects ? Tell him that some adventurers are engaged upon a steamboat at Leghorn, to make the *trajet* we projected. I hope he is charitable enough to pray that they may succeed better than we did.

Remember me most affectionately to Mrs. Gisborne, to whom, as well as to yourself, I consider that this letter is written. How is she, and how are you all in health ? And pray tell me, what are your plans of life, and how Henry succeeds, and whether he is married or not ? How can I send you such small sums as you may want for postages, &c., for I do not mean to tax with my unreasonable letters both your purse and your patience ? We go this summer to Spezzia ; but direct as ever to Pisa,—Mrs. —— will forward our letters. If you see anything which you think would particularly interest me, pray make Ollier pay for sending it out by post. Give my best and affectionate regards to H——, to whom I do not write at present, imagining that you will give him a piece of this letter.

Ever most faithfully yours,

P. B. S.

# INDEX